"Well done and thank you Dr. Joe Way! W
need help in utilizing the modern tools of
the great commission. Ultimately, the Holy
hearts toward Christ and with each generation there are new ways to see
God's heart. In *Producing Worship: A Theology of Church Technical Arts*, Dr. Way gives
every leader, pastor, and tech team vital insights on how to maximize the potential
of new technology to help transform and equip the Church to proclaim the timeless
message of Jesus Christ."

~ Paul Baloche, Dove Award Winning Singer, Songwriter, and Worship Leader

"In an age of technology and increased digital engagement in the Church and the
world, Dr. Way provides a helpful and hopeful theological framework for understanding and using technology in the worship of our Lord. Dr. Way's book is a must
read for all those who desire to lead faithfully God's people in worship. It is a com-
pelling, original work from a scholar with great biblical and theological acumen, as
well as an expertise in producing worship."

~ Dr. James Spencer, Theologian and Vice President, The Moody Center

"We don't often think of modern technology as being part of what happened in the
early Church. It has always been there, from the design of buildings, to works of art
that express worship. This book is so much more than just a book about the use of
today's technology in worship. It is a deep dive into how 'technology' has always
been used since before the time of Jesus. Joe has expressed how today's technology
is the next logical step in using what we know today in order to help people find
Jesus. A great explanation the true purpose for technology and how we arrived to
where we are today, using it for the same purpose that has lasted for thousands of
years: Reach people for Jesus, using everything we have to offer."

~ Greg Baker, Technical Director, Saddleback Church

"I can think of few things that people complain and divide over more than the 'production' of a church. Sound. Lighting. Song selection. Instrumentation. All of it. And
our preferences in these areas tend to rule in our hearts as to whether Jesus is still
worthy of worship, based on whether or not the church service fits our preferences.
Yet all the while, the perfect, flawless, matchlessly glorious Savior stands ever worthy
of the greatest praise we could give and more. I am grateful that Josiah Way has
written this resource to equip the church in this critical area that so few have written
on. I pray it will spur on conversations that end in God getting the glory he deserves."

~ Stephen Miller, Pastor and Author, *Worship Leaders, We Are Not Rock Stars*

"With technology advancing rapidly, it's all too easy for church technical artists to
get caught up with all the gadgets and the technicalities and forget the spirituality.
Joe's really important study, *Producing Worship*, engages this challenge with enthusi-
asm, insight and theological rigour, offering a helpful new perspective on the role of
the technician in worship which will be an inspiration to many."

~ Dr Andrew Davies, Reader and Director, Edward Cadbury Centre for the
    Public Understanding of Religion, University of Birmingham UK

"Joe Way's new book, *Producing Worship*, peels the onion of the church technical artist in ways I had not really thought of. As someone who served in full time ministry for twenty years, I love the way this book lays out what we do and why we do it in solid detail. It also gives validation and a Biblical footing to what church technical artist accomplish every week. Most of all I love the heart of this book. Joe truly loves those who serve the church and accomplishes the fine art of encouraging the ones in the back of the room, in the dark. I challenge not only technical artists to read this book, but Pastors, and senior leadership as well."

~ Van Metschke, Church Relations, CCI Solutions; Co-host and Founder, *Church Tech Weekly*, Church Tech Arts

"*Producing Worship* is a must read for pastors, tech directors, and anyone serious about working in the church technical arts. Church tech production is the most unique ministry and requires the highest level of craftsmanship—where scientific knowledge, technical know-how, and independent creativity are equally important. It's also a ministry that confuses a lot of people, from pastoral staff to congregants to even those within the tech arts community, because they all struggle with the same question. This question is one Josiah has tackled and answered; 'How does God want the tech arts used in my church?'"

~ Chris Huff, Author, *Audio Essentials for Church Sound*; Host, *Behind the Mixer*

"As those who place their hands to the sacred duty of helping create space for many to worship through songs, creative elements, and experiences, this book helps lift our eyes to the holiness within what we get to do as church technical and production teams by bringing a fresh theological perspective on what the Bible teaches around what we see today in our production and technical services."

~ Daniel Scotti, Central Production Director, Saddleback Church

"Though we may not like the sound of it – a Worship Experience is a production. It is an art to take an audience on a journey from beginning to end. Dr. Way beautifully captures the biblical foundation for worship and details how the technical arts fit within the context of the modern church. A rare foundational look at the inner workings of what it takes to craft a dynamic, engaging Worship Experience. *Producing Worship* is a must read for any creative church staff or volunteer."

~ Carl Barnhill, Owner, Twelve:Thirty Media; Host, *The Church Media Podcast*

"If you have served in a church tech ministry for any length of time, you know Sundays are more than PowerPoint, cables, or social media posts. Joe not only gives a specific understanding of how church tech is rooted in Scripture, but furthermore gives a Biblical model of how we can serve by honoring God with the mediums we produce onstage, through the camera lens, and produce worship that may win the hearts of those we reach. For anyone who is wanting to do more than show up for church tech, this is a book you must explore."

~ Jeremy Smith, Author, *Rebuilding*; Host, *ChurchMag* Podcast

"Something that all church staff and worship/tech volunteers should read."

~ David Jordan, Creator, Church Sound & Media Techs Conference

"This is a splendid book filled with wisdom, insight, and experience that every church can use."

~ Quentin Schultze, Ph.D., Author and Theologian, *High-Tech Worship?*

"This is 🔥🔥🔥! It must be published! Josiah Way's work has shown that the church audiovisual technician is more than just the person who runs the slides on Sunday morning. Using sound exegesis, Way shows us how this person is an artist created in God's own image and imbued with God's wisdom to develop a craft for the glory of God and good of his people. Way's work should usher in a new emphasis on the AV technician as an artist and core aspect of the worship experience."

~ Russell L. Meek, Ph.D., Assistant Professor of Old Testament and Hebrew, Louisiana College

"Love. Love. Love. As someone who's worked in church production for 20+ years, I'm so excited to finally see this discussion being brought forward, and in such a spectacular way. Joe has put together a well-reasoned and biblically-sound correlation likening church technical artists to the skilled craftsmen as referenced in the Old Testament, using their very best to help communicate and teach the message of Christ through the use of technology. Thank you for putting a voice to many of the thoughts I've had in my time in ministry. I want my whole team to read this book."

~ Rob Mortenson, Production Pastor, Saddleback Church

"I am excited that Dr. Way has put into words what many leaders and artists in the church need to hear. Since the Church has been called to be salt and light in this age of unprecedented technological development, I believe that the need for this book is long overdue! Pastors, leaders, and those serving in the technical trenches will greatly benefit through devouring this content. *Producing Worship* is not only instructional, it will also bring encouragement and focus to the calling of the technical artist."

~ Chaz Celaya, Professor and Director of Commercial Music, Point Loma Nazarene University; Author, *The Sound Guide*

"The use of technology has always been a part of those on the cutting edge of worship. Joe's book helps us see this and expands our vision of worship for the church."

~ Phil Thompson, StreamingChurch.TV

"Having spent many years in the tech industry reading manuals, learning technology, and producing shows I have never run across a book or seminar that challenges technical artists to look deeper into why they do what they do and how they fit into the mold of the larger Church. Because of this, I have seen a lot of burnout among techs. Joe Way's book, *Producing Worship*, is a necessary look into the how and why of church technology through an exegetically biblical lens. This book is a great resource for the church whether you're a tech, a pastor, a music director, or just another attendee with questions about the growing technology in church."

~ Randy Murphey, Manager of Multimedia Services, California Baptist University

Producing Worship
A Theology of Church Technical Arts

JOSIAH WAY, PHD, CTS

All Scripture quotations, unless stated overwise, are from *The Holy Bible, English Standard Version* (ESV). Copyright © 2001 by Crossway, a publishing ministry of Good News Publishers.

Working translations from chapters 3–5 are derived by the author from both the original languages and eight most popular English translations including: ESV, NRSV, NASB, NET, NIV, NKJV, NLT, and CSB.

ISBN-10: 0578-41918-1
ISBN-13: 987-0-578-41918-3

Library of Congress Control Number: 2018913778

Front cover image: Josiah Way.
Book design: Josiah Way.

First printing edition 2018.

Published by: Josiah Way, Lake Forest, CA.

www.josiahway.com
www.producingworship.com

Printed in the United States of America.

DEDICATION

To those who selflessly serve their local church
through the technical arts ministry.

CONTENTS

FOREWORD

I wanted to title this foreword: "Why I committed academic suicide on my first book." After spending four years of PhD study and two years of Master's work, the next step—I was told—was to "find a legitimate publisher" to disseminate the doctoral thesis as a manuscript. Otherwise, your work will not be "respected." And I admit, having a published academic work has always been a dream of mine. So, I have to first say to the two traditional academic publishers who accepted my thesis and were willing to "sign me" and my crazy idea of publishing a theology of church technical arts, thank you. I am honored and humbled. But I decline.

So why would I turn down the opportunity to make a dream come true? To steal the opening line from *The Purpose Driven Life*, the #1 bestselling book of all time after the Bible, because it is not about me. It never was. It is about God and one another.

Technical artists are dear to my heart; I have a deep respect for those who dedicate their lives to serving their local churches, often doing so from the dredges of the church, back in the corner, stuck in the dark. They show up before everyone else, and are often the last to leave. When the job is done well, they go unnoticed. And often un-thanked. But they do it because God has instilled in them a unique talent and shaped them to serve in that capacity.

The initial purpose of this research was to explore what the Bible says about church technical arts and to answer the question: how might a biblically informed theological understanding help better shape praxis for contemporary church technical arts? The pursuit of knowledge was an important journey, and it resulted in the everlasting "doctor" title. However, knowledge is useless if not shared. I committed academic suicide in order to build a bridge between the theological study of technical worship and the practical application of it. This scholarship needed to get into the hands of church techs, pastors, worship leaders, scholars, academic institutions, church creatives, and lay servants. As a new author, the price point would have been prohibitive for mass distribution. While my dream would have been reached, the purpose would not have.

I look forward to this book informing the theological choices we make in the application of technically generated sound, light, and visuals within our worship services, as well as being able to set the table for church leaders and their creative staffs to better serve God, one another, and their congregations.

ACKNOWLEDGMENTS

This book would not be possible without so many people believing in the purpose. First and foremost, that is my wife Amy who took on all the extra load while I accomplished this goal. For that I am forever grateful. To my kids Kat, Morgan, and Tyler, let this be an example that anything you wish to achieve, at any point in your life, is possible if you put your mind to it.

To my doctoral supervisors, David Cheetham and Andrew Davies, professors James Spencer, Russell Meek, Alin Vrancila, J. Brian Tucker, Thomas Habinek, Kevin Robb, Greg Thalmann, and Dallas Willard, and classmates and colleagues who helped shape me into a critical theologian, you have challenged my beliefs and honed my methodologies. I am a scholar because of you all. To my fellow church techs and AV pros, this book is due to the inspiration that came from serving in the trenches with you day in and day out. If I named you all, this entire book would be nothing but a list of names. Nevertheless, I want to give special recognition to my Saddleback Production volunteers, my fellow Saddleback technical directors and worship staff, my original Compass Tech team where it all began, and my Cal Baptist student workers and fellow AV staff. You are all true examples of servant leadership. Last, I want to thank my Savior for gifting me with the ability to gather and share this research knowledge and allowing me to serve your church every day in "producing worship."

๑ 1 ๔

SETTING THE STAGE

INTRODUCTION

In 1985, renowned physicist Freeman Dyson cited technology as the greatest gift from God after life itself. At that point the first Macintosh personal computer was only one-year old, and Windows 1.0 would be released just a few months later. Reflecting on the power of nuclear energy and basic flow of information, Dyson concluded that it takes fifty to one hundred years for a new scientific technology to catch on and that even though technology is often thought to be growing more rapidly than in former times, this is an illusion caused by our current perspective. Undoubtedly, the invention of the microchip changed that forecast: technology today is advancing exponentially—doubling nearly every two years according to Moore's Law. It is no surprise then that as a society just over thirty years removed from Dyson's proclamation that record rates of technological consumption have resulted in generational segments that are now more often defined by technical innovation than decades or political movements. As a result, the contemporary church, which remained relatively stagnant in its liturgical worship methods for centuries at a time, has had to answer to how it will embrace modern technological advances. Quentin Schultze suggests that "the idea of technological adaptation brings us to the biblical basis for human use of technology, namely, our calling as caretakers or stewards of God's creation."[1] This technological calling has historically been overlooked in both contemporary scholarly research and everyday church practice.

The product of a skilled technical artist can present the biblical narrative in the same imaginative and meaningful way as other great works of art

---

[1] Schultze, *High-Tech Worship*, 46.

utilized by the church throughout history. Just as an artist once realized the telling of God's story could be enhanced through converting a standard window into stained glass, Michelangelo forever impressed the creation of Adam onto humanity through a paint-brushed fresco upon the ceilings of the Sistine Chapel. David's physique is immortalized in our modern minds through stone. So, too, the method by which the church would deliver its message forever changed when Emile Berliner combined his microphone invention with a modified Thomas Edison phonograph to create the first usable tool for recording and playback of sound to large groups of people at long distances simultaneously. When C. W. Rice and E. W. Kellogg patented the first loudspeaker in 1925, technology entered the realm of everyday public speaking and, in turn, the pulpit and pew. Church art became a technological endeavor. Over the next forty years, as affordable lighting and video capabilities developed for the general public, the church sanctuary became a living theatre for God. To the chagrin of some, sanctuary projection screens have become the cross and stained glass for the electronic age, constantly creating and forming religious imagery that was previously represented by architecture and icons. And thus, high-tech, digital liturgy is now the contemporary church's art form and vehicle for presenting the gospel message to a technological generation.

Even with nine decades of technological assistance in presenting the pastoral message, the study into the theology of church technology has been glaringly absent. Thus, this book moves us towards a theology of church technical arts and the practice of "producing worship." It performs a theological examination of church technical arts, developing a technological metanarrative through three key relevant texts: Exodus 35:30–36:1; Hebrews 2:12–13; and Colossians 3:16. In practice the texts can be read to view technical artists as serving as Christ's mediators between the stage and congregation (Heb 2:12–13) for the purpose of building up one another in the church through multi-medium worship (Col 3:16) and that technical artists are defined by artistic excellence through craftsmanship, skill, creativity inspired by the Spirit, and performance through a heart of service possessing the characteristics of intelligence, wisdom, knowledge, and the ability to teach their craft to others (Exod 35:30–36:1). The characteristics of ability, artistry, wisdom, craftsmanship, and technology are demonstrated through the Old Testament artisans Bezalel and Oholiab and the construction of the tabernacle. The New Testament verses of Hebrews 2:12–13 and Colossians 3:16 first clarify the use of technology in contemporary worship practices by presenting Jesus as the perfect intercessor between the worshipers and worshiped and, second, by offering the purpose for how the church should institute technological worship, namely, the sanctification and building up of the congregation. Last, the findings are viewed in light of current practice. The view presented formulates a portrait for contemporary sound, lighting, and visual techniques, as well as the technical artists' purpose and position within the church. In this way, the technical arts become representative of larger theological principles when engaged within the church context.

This book forms a portrait of the church technical artist. It is not for those looking for a light work of praise or excuse. It is an exegetical work exploring the biblical text with practical application. Chapter one is contextual; chapter two historical; three, four, and five are textual, and chapter six is practical. Each chapter builds one pillar of the theological tower that is formed in chapter six. As a reader, be prepared to explore the texts as would biblical scholars. The remainder of chapter 1 establishes the purpose and aim and its relevance to the modern church. Working definitions of relevant terms are presented, along with explanation of the various research methodologies utilized.

Chapter 2 answers the question, "How did we get here?" How contemporary church technical arts developed out of the church's historical understanding of musical worship is unpacked. The major turning points in history from the Old Testament period, the Reformation and Enlightenment, and the recent "worship war" debate in Contemporary Christian Music (CCM) are outlined. Since the NT church is fundamentally a people rather than an organization or infrastructure, essential to understanding how the technical arts apply to contemporary church practice, an analysis of the various mediums for biblical and practical instruction that directly impact the thinking, beliefs, values, and practices of modern church technical artists is performed.

Chapter 3 explores how technical artists are to perform their craft as viewed through Bezalel's and Oholiab's call to construct the tabernacle and its holy items in Exodus 35:30–36:1. Moses steps aside and entrusts these artisans who are called to their craft and filled with the Spirit in all skill, wisdom, knowledge, craftsmanship, and teaching. Exodus 35 details the care God's technical workers should possess when performing work for and in the sanctuary. The view presented becomes a model for modern technical artisans.

Chapter 4 unpacks Hebrews 2:12–13. In this passage Christ is the Singing Savior, mediator between God and humanity. Jesus's role serves as a contemporary metaphor for church technical arts. Just as the Son intercedes between the Father and believers, so too technical artists are physical mediators between the stage and congregation. Likewise, technical artists function as the church's mediators of the worship response to Christ.

Chapter 5 explores the fundamental meaning of "psalms, hymns, and spiritual songs" in relation to the surrounding passage as presented in Colossians 3:16 and its cross-reference Ephesians 5:19. This chapter answers the why question. Why do technical artists serve as God's church mediators? In order to teach and admonish the body of Christ for building up one another, Paul exhorts the church body to sing psalms, hymns, and spiritual songs.

Where worship of God is performed vertically, the tangible effect is a horizontal directive that sanctifies the corporate body.

Chapter 6 consolidates chapters 3–5 and explores how the scriptural foundations work together to form a biblically based understanding of the technical arts in practice; a technical arts metanarrative for "producing worship" is fashioned out of these findings. The book and chapter conclude with suggestions for future theological research and discovery.

Chapter 7 offers a final thought and challenge to church techs.

WHAT IS "CHURCH TECHNICAL ARTS"?

Church technical arts—also known as church production—extends beyond traditional worship-band or chorale-led congregational singing to include the latest in specialized, contemporary production methods. Examples include multimedia presentations, amplified sound, digital audio workstations (DAW), LED concert lighting, large screen image magnification (IMAG), digital effects and enhancement, live and on-demand video streaming, theatrical set design, assimilation of social media, professional graphic design, stage hands, and production staff. These various methods for presenting the biblical narrative to the congregation are the current expectation, not the exception. John Dyer of Dallas Theological Seminary notes that to provide an attractive worship service for the millennial generation, church leaders commonly consider technology a "necessary evil that we use until Christ returns."[2] Even though most modern churches do accept technological worship as the new norm, some still hold reservations. Kevin Bergeson, for example, suggests that it is the biblical text itself that should force the congregation to utilize their imaginations, and when fed digital photos and videos, congregants mistakenly assume that the visuals are the actual events and period loci instead of allowing the Holy Spirit to work within them.[3] Nevertheless, most churches today are keenly aware of the positive aspects in attracting the younger generations, as well as the added benefits of technological enhancement, like the ability to reach are larger audience by taking services

---

[2] Kevin Hendricks, "John Dyer: Technology & the Church," *ChurchLeaders.com*, December 14, 2011, no pagination, http://www.churchleaders.com/worship/worship-articles/157116-john-dyer-technology-the-church.html.

[3] Bergeson, "Sanctuary as Cinema," 303. Bergeson uses the example of showing a picture of the Sea of Galilee today to illustrate what the area would have looked like during Christ's day. In doing so, he suggests that the church could be giving a false impression to the congregation. Even though, his warning is noteworthy, Bergeson's suggestion is problematic. Well-produced graphics, sounds, and videos do have the ability to stimulate the imagination as well—as will be demonstrated later—translating the text in the way the preacher intends to present it. Nevertheless, it is significant to recognize that the use of technical productions in worship services is not fully accepted by all, even if it is the norm in contemporary church settings.

online. The necessity for producing technological worship to be deemed relevant has led to a new phenomenon in the past decade: the technical arts ministry. The influx of new church technical artists has spawned regional, national, and worldwide resources and organizations that focus on training church technical artists in praxis and purpose.

THE REASONS AND AIM FOR THIS BOOK

Technical artists serve a fundamental role in the modern church, a role that is not solely a job but also a source of producing worship itself. Prior to Saddleback Church, I worked as the technical director a nondenominational, evangelical church that prioritizes the integration of emerging technologies with Christian contemporary rock-style worship, Compass Bible Church in Aliso Viejo, CA. At a mid-week rehearsal, an unknowing guest musician asked the front-of-house audio engineer what instrument he played. "I play the sound board," Jon Lillie responded. In those five words Lillie summed up the truth that technology in the contemporary church is not comprised solely of pieces of equipment that allow the worship band be seen and heard but is an instrument itself. To serve its designed purpose, technology must be played, mastered, and utilized as a tool for the service of the church and the church body. This predicates that there is also as much responsibility for the artists operating the equipment within the church context as there would be for any other worship-ministry musician. Likewise, during the weekend services my volunteers would often ask me to clarify statements the senior pastor preached during the sermon. To do so meant that as a tech leader, I had to be as aware of the theological content as much and the visuals and audio.

Even though technology in the modern church is becoming commonplace and the use of the arts in church ritual is firmly rooted in the biblical text, those assigned to employing the arts for the modern church often have little or no formal training, either practical or theological. Technical artists are frequently thought to be the guys who constantly complain about the hard labor and long ministry hours, ask for exorbitant budgets for new "toys" in the form of expensive audiovisual equipment, and routinely voice why some new idea just will not work. Why such negativity toward the church's modern artists and their artistic medium of technology? The evidence appears to suggest that it has developed out of a lack of understanding by both the church and the technical artists themselves of their functional and theological roles in the church. It is this disunion that necessitated the present study.

Three main objectives are being sought: (1) initiate the scholarly discussion on what I call "producing worship," utilizing technology in the performance of the worship service; (2) discern how the Bible speaks of those charged with producing technological worship services; and (3) discover how technical artists might apply the findings to praxis in terms of both spiritual development and daily operations. Due to the speed of technological change, the way technology "looks" just a decade from now will be drastically different than today. Therefore, focus is placed on current practices rather than specific equipment. Because the Bible does not directly speak to technology in the way it is thought of today, the text will be utilized as a tool to shine light on praxis. Placing a firm "flag in the ground" is the first and foremost goal. The mere fact that this book exists accomplishes that objective. Filling the disciplinary gap between practice and biblically informed theology for church technical artists—an overlooked church demographic—is this book's most significant accomplishment. This book will not make judgements about what technical artists ought to do or ought not to do when it comes to everyday audio, video, lighting, and scenic choices. It will not directly state which practices are better or more effective than others. Yet, it will inform how the role of the technical artist is viewed both scripturally and historically. It is this knowledge which should then inform appropriate choices for praxis.

This book performs a constructive theology: when technical artists perform their craft from within the church context, larger theological principles are at work. Church technical artists are modern-day characters in God's plan within the greater theological school of ecclesiology. This book uses the Bible to demonstrate what those larger principles may be and how technical artists can apply those principles to praxis. It holds to Kevin Vanhoozer's theological concept of "theodrama," wherein everything God does in and through the Bible is part of a greater metanarrative that plays out God's ultimate design for humanity. The actions of every biblical character are an essential aspect of God's dramatic work in which God is the primary director of the actors. In this way, Christians today are fulfilling their roles in this epic drama, which continues beyond the biblical text. The Bible is used to offer insight from past actions and characters—like those of Bezalel and Oholiab—to speak to contemporary practice. Even though the original characters in the biblical theodrama could not have possibly imagined the technical arts as used in the church today, parallels can be drawn from the exegetical findings and, therefore, contemporary functional applications.

Biblical narratives are used retrospectively, not as a precursor as if they were originally intended to speak to today's actions, but from a modern understanding looking backward, drawing conclusions that the original authors or recipients would not necessarily have comprehended, while at the same time maintaining the text's implicit doctrinal stance. Andreas Köstenberger and Richard Patterson refer to this as the "hermeneutical triad," which is a holistic approach to the Bible that balances history, literature, and theology, wherein "rather than being pitted against one another . . . each have a vital place in the study [of the Bible]."[4] In order to accomplish this, they suggest, it takes balancing a historical-cultural awareness, an understanding of the passage's place in the canon, examining the literary genre as a tool for presenting the biblical theology, grasping the text's linguistic usage, and finally being able to apply the theology to life and practice. Grant Osborne adds that finding meaning in a text has "three foci: the author, the text, and the reader. The author 'produces' a text while a reader 'studies' a text."[5] Therefore, I propose that once the text is written and the author has passed, the text must become autonomous, carrying a transferred meaning as perceived from the loci of the reader, now two to three thousand years removed from the original. In this way, the text speaks to praxis as the application of earlier principles. Contemporary practice is not inserted back into the text; readers balance a perceived understanding of historical context, intended meaning, literary style, and theological doctrine.

This study asks the Bible questions it is not seeking to answer. There is no reference to modern technology in the Bible. This book asks the Bible to shed light on current practice, specifically for individuals charged with running the technology for worship services. In order to accomplish this, theological perspectives regarding church technical arts are drawn from a thorough exegesis of three distinct scriptural references: (1) Exodus 35:30–36:1: becoming skilled craftsmen, wise in practice; (2) Hebrews 2:12–13: Christ the mediator between God and humanity as a metaphor for church technical artistry; and (3) Colossians 3:16: the purpose of multi-medium worship for building up the church and one another. The Bible is explored according to the likely reception of the text in its presumed original context via the scholarly discussion in modern theology, concluding with applying the exegetical insights gained to contemporary practice.

DO THE TECHNICAL ARTS FIT INTO THE CHURCH SETTING?

Until recently, the legitimacy of the technical arts was relegated to the "worship war" discussions of style versus substance, dictated by worship leaders and pastoral leadership for whom technology is a sister-tool of the trade, not a theological ministry in which they are personally engaged. To this end, technical artists have been playing a game of catch-up. While surveying the popular print and online trade publications, a possible explanation is discovered: while approximately three-fourths of every worship-leader-type magazine focused on scriptural priority, with minimal selections on instruments, gear, and other tangible necessities, the production-type resources

---

[4] Köstenberger and Patterson, *Invitation to Biblical Interpretation*, 78.
[5] Osborne, *The Hermeneutical Spiral*, 465.

included zero total articles over the span of the previous six years that explored the craft with a traditional, exegetical methodology. The majority of the material in technical arts trade resources is aimed at exhibiting new gear, tutorials, devotionals,[6] and the occasional "leadership" or team building, "you-can-do-it" pep talk, with miscellaneous scriptural references sprinkled in so to sound biblical. The analysis of my observation is not meant to challenge either the motivations or validly of the biblical presentations that do exist. Instead, I hypothesize that the reason may lie in the fact that technical artists appear to use Scripture to justify their actions, whereas generally biblical scholars would prioritize the Bible, allowing it to shape praxis.

John Dyer suggests a fundamental quality concerning the technical arts in practice:

> As technology creators and workers (web designers, videographers, sound techs, etc.) . . . people care more about what you [say] . . . than what their pastor says. Right or wrong, people don't care their pastor doesn't have a Facebook account or if he likes to memorize Scripture instead of just accessing it on his phone. But if you—the tech guy or gal—does or says something unexpected about technology, people notice. You have a voice and a position of authority, so use it wisely.[7]

Congregations are indeed watching. Technical artists' actions effect what others believe about the craft—and ultimately about Jesus—when performed within the church setting. The practice of technical artistry produces theological worship; what technical artists do says something about the church's beliefs. Thus, as ministry leaders and representatives of their respective churches, technical artists share a greater responsibility than simply leading their teams in excellent technical production; they also influence spiritual development as functional theologians themselves. Ultimately, then, it could be suggested that the theology practiced and proclaimed from the production rooms is equally shared by both the techs themselves and the person at the other end of the microphone.

In a 2013 *Church Production Magazine* online article, Mike Sessler, co-host of the industry-popular podcast, *Church Tech Weekly*, poses the question: "What would happen if the tech department became known as the most spiritual department in the church? It's a valid question."[8] Indeed, it is. Sessler answers this himself: "As much as I want to be known for being an excellent technical artist, I think I would rather be known as someone whom God has

---

[6] Even though devotional style writings do exist and may be scripturally based, they are not theological in a traditional academic sense. They are thematic interpretations of the Bible used to support a presupposed theme, rather than directing the theological stance. For this reason, they are not considered a textual exegesis intended to inform theological understanding.

[7] Hendricks, "John Dyer," no pagination.

[8] Mike Sessler, "Cultivating the Spiritual Side," *Church Production Magazine*, 2013, no pagination, https://churchproduction.com/education/cultivating_the_spiritual_side_of_being_a_church_tech_director.

filled with His Spirit." This leads to the question, how would the church, pastors, ministry leaders, the congregation, volunteer teams, and ultimately Christ himself view technical artists if technical artists truly presented a biblical focus in both life *and* practice? Could the church's assimilation of the technical arts into the worship services rely as much on the people running the service as it does the technology itself? Is, then, a proper theological understanding of the role essential to performing that role suitably?

OVERVIEW OF GOSPEL CENTERED TECHNICAL ARTS

Acts of building and creating in the Bible are plentiful. Craftsmen are used to construct many great symbols of God's dwelling and favor, such as Noah's Ark, the tabernacle, and Solomon's Temple. Yet, craftsmen also create symbols of humanity's desire to control God through earthly creations like the golden calf and Tower of Babel. It is not that God called craftsmen to build and labor for his kingdom or that "Jesus was a carpenter" that is noteworthy. What is significant is that the Bible defines specific qualities, aligned with a specific purpose, that the church's craftspeople ought to possess while performing their craft. The sign of a true craftsman is that he or she is obsessed with the details that work together to form the final product.[9] A true biblical craftsman, then, would view the purpose behind the final product as a praise to God while building up one another, manifested through the person of Christ and inspiration of the Spirit.

The key passages explored here—Exodus 35:30–36:1, Hebrews 2:12–13, and Colossians 3:16—work together to demonstrate that Christ is not only the object of worship, but that it is through Christ that worship is offered, guiding the artist's skills through wisdom and knowledge of craft to form the details for presenting God to the congregation in order to instruct the corporate body. The application of church technical arts must be viewed through the lens of a Christian metanarrative that is faithful to the gospel because Scripture dictates that all Christians do—including their artistic endeavors—must be done as for God (Col 3:23; 1 Cor 10:31).[10] "Being biblical is thus a matter not only of theory but also of practice," suggests Kevin Vanhoozer.[11]

---

[9] Nicolosi, "The Artist," 116.

[10] For the sake of the book the Christian gospel is understood as such: It is known that due to the sin problem, humanity was eternally separated from a holy God, and the only solution for reconciliation was for the Father to send his only Son to live a perfect life, take the wrath humans deserved, being tortured and crucified on a Roman cross, and raised again in three days in accordance with the Scriptures, overcoming death, so that he would stand as the propitiation before the Father, so that those who believe may be seen spotless, blameless, and forever reconciled; yet until each man or woman's earthly deaths or Christ's return, followers are to obey his commandments as doers of the Word, bear the fruit of righteousness, love God and one another, and become further conformed into the image of Christ Jesus.

[11] Vanhoozer, *Faith Speaking Understanding*, 1–2.

How technical artists execute their role speaks to what they truly believe about Christ. To perform their craft without recognizing who God is, what he has done in reconciliation, and how he calls believers to live would be to perform it with no higher value than held by the secular world.

For believers, Scripture reveals the entire biblical narrative. Therefore, technical worship in practice must be performed in a way that conforms to that end. Philip Stolzfus suggests that Christian art must be grounded in a wholly biblical doctrine to avoid "potentially idolatrous capabilities of the artist. The artist should be viewed as a craftsman, utilizing fittingness as the ultimate criterion in correlating the given material of an art form with the responsibility and obligation of the artist."[12] John Witvliet adds that a biblically informed understanding "challenges us to embrace the whole gospel rather than just one part of it . . . [which] can be powerfully life-giving, showing us aspects of the gospel of Jesus that can transform, deepen, and encourage . . . [believers] for faithful ministry."[13] That is the claim this book makes: there are truths regarding technical worship buried within the gospel narrative. In this way, technical artists are to approach their craft holistically, not focusing on any one aspect of either their craft or Christian life but rather appropriately encompassing the faith in its entirety.

Of course, one cannot simply open the Bible and read what it says about the technical arts. Yet, both the biblical narrative and church history can be viewed so to suggest that God has endowed all artists—beings created in his image—to be used for his purposes. To bridge the gap between traditional arts and the technical arts, Tim Keller contends, the church must "fill in the blanks that the Bible leaves open,"[14] drawing conclusions from parallel examples and allowing them to inform modern practice. In this same way, Clayton Schmit suggests, "any art or technology that truly advances the promise of the gospel or successfully draws people into an encounter with God is worth exploiting"[15] and exploring. Therefore, to evangelize a culture, technological worship must contextualize the gospel's ministerial loci, their individual congregational sphere of influence. Artists must "speak" the gospel through their creations in a way that is relevant to their cultural context. In today's ethos that means through technological mediums.

Technology is God's modern voice. Therefore, this book argues that technical artists have a responsibility to fully understand their specific role in God's dialogue with his people. Technical artists serve as God's mediators between the pulpit and congregation, purposed with the task of leading the congregation in vertical praise to God and horizontal edification of one

---

12 Stolzfus, *Theology as Performance*, 8–9.
13 John Witvliet, "The Joy of Christ-centered, Trinitarian Worship," *Worship Leader*, April 14, 2015, no pagination, https://worshipleader.com/articles/the-joy-of-christ-centered-trinitarian-worship/.
14 Keller, "Reformed Worship in the Global City," 198.
15 Schmit, "Technology and Art in Worship and Preaching," 41.

another. Hebrews 2:12–13 suggests that just as Christ is the mediator between the Father and humans, so too in the church context technical artists are tangible mediators between what is presented on stage through the pastor and worship teams and the audience. Through physical equipment the message passes; technical artists translate and present that message. Performance impacts response. The message presented—for better or worse—necessitates a responsibility in craft that extends beyond knowledge of the discipline to a full recognition of the role as facilitator between the stage, the congregation, and Christ. Therefore, this study fills the "blind spots . . . that involve digital media" which have been left out of religious studies and are "only now coming more fully to the attention of scholars" as noted by Frank Burch Brown's most recent and relevant work on the discipline, *The Oxford Handbook of Religion and the Arts*.[16] Even though Brown cited this gap in the scholarly discussion in 2014, not until this study has there been an attempt to fill the lacuna. It is this intersection between practice and theology that this analysis seeks to inform.

DEFINITIONS

In order to properly set the stage, working definitions for common terms need to first be established.

Church Technical Artist. Throughout the scholarly literature, no specific definition of technical artists—also known as "techs" or "creatives"—surfaced, yet technical artists serve a specific and important role in producing and administering the church's worship service. As chapter 4 will demonstrate, just as Hebrews 2:12–13 presents Christ as the mediator between the church and the Father, so too technical artists serve a tangible role mediating the presentation of the message of Christ from the pulpit to the congregation. Therefore, as a working definition, church technical artists are those who employ electronic mediums in the production of a worship service through a learned skill set in a performance-based creative capacity. Three central concepts surface out of this definition. First is the idea of craftsmanship. The technical artist is one whose final product is created through a purposeful utilization of specific tools of the trade and learned specialized knowledge. Second is aesthetic creativity. The Christian technical artist utilizes his or her art form as a mode of worship, inspired by and through the Holy Spirit. Third, the technical arts are an action-based, performed theology aligned with a heart conformed to both Christ and the congregation to build up the church body. Lee Bloch suggests that technical artists are "one part engineer and two

---

[16] Brown, "Introduction," 5.

parts artist . . . [and] to further fully understand the technical artist, you need to realize that you are dealing with a person who is driven by a vision, an image, of how the end result should look or sound."[17] It is a vocation ingrained in the liturgical structure of the modern church. Those assigned to serve the artistic production needs of the worship service are placed in God's order for his specific purpose, to fulfill a calling that is one's vocation.

Craftsmanship. The *Concise Oxford English Dictionary* defines a craftsman as "a worker skilled in a particular craft," with "craft" being "an activity involving skill in making things by hand," as well as the "skill in carrying out one's work" and "members of a skilled profession."[18] Applying this to a religious context, the *Dictionary of Bible Themes* outlines "craftsmen" as "skilled workers and artisans. God is often compared to a master craftsman in his work of creation. Scripture also identifies God as the giver of talent and ability to certain individuals, in particular, those employed in the construction of the tabernacle and the temple."[19] Therefore, for the purpose of this book, biblically rooted craftsmanship is defined as the professional work performed by a skilled artisan gifted with the ability to create out of his or her own God-granted abilities with the purpose of pointing his or her audience to the object of the creation. In terms of Christian craftsmanship, the object is the trinitarian God revealed through the person of Jesus Christ.

Aesthetics. According to the *Concise Oxford English Dictionary*, aesthetics is "a set of principles concerned with the nature and appreciation of beauty, especially in art. [It is] the branch of philosophy which deals with questions of beauty and artistic taste."[20] *Dictionary of Bible Themes* adds that biblical beauty is "a physical or spiritual quality which brings pleasure to those who behold it. Scripture stresses the beauty of God himself and his creation while noting that the beauty of the creation can lead away from God or become idolized."[21] Likewise, aesthetic language is frequently associated with the glory of God (Exod 28:2b, 40b). Scott Aniol writes:

> The Bible is filled with aesthetic terminology to describe God. God's glory is his beauty, and his beauty is magnified when his people delight in lesser forms of beauty. In the Bible, beautiful music is often used as a way to magnify and praise the beauty of God himself.[22]

Glory is a distinctive characteristic relating to God's presence, identified through his work in creation, and known by his visible (brightness/light) and active (powerful) manifestation. For believers, God's glory became supremely magnified in creation through the life, death, and resurrection of

---

[17] Bloch, *Worship from Backstage*, 219–28.
[18] Soanes and Stevenson, *Concise Oxford English Dictionary*, 333.
[19] Manser, *Dictionary of Bible Themes*, 5272.
[20] Soanes and Stevenson, *Concise Oxford English Dictionary*, 21.
[21] Manser, *Dictionary of Bible Themes*, 4040.
[22] Aniol, *Sound Worship*, 8.

Jesus Christ. Therefore, aesthetics within church technical arts is the beauty found within a created product that brings glory to God and extols the characteristics of him. Its qualities will lead God's people to seek beauty in God's work and identity.

Technology. Today, technology conjures up images of electronic devices, social media, and global communication. Yet, the term's etymology descends from the Greek term *tekhnê* which has the direct meaning of craft, trade, and art, and from where we get the concept of craftsmanship. Aristotle alleged that *tekhnê* has the sense of possessing the skill of rhetoric; that which is created has the capacity for persuasion and the power to influence thoughts and actions. Therefore, all technology influences the reality experienced by the congregation through the way it is integrated. In this way, technology is fourfold: (1) it is the physical equipment used in service of God; (2) the meanings that we attach to the equipment; (3) the way the equipment is used; and (4) the meaning the congregation perceives from the final product, which may or may not be the meaning intended by the technical artist. Each of these details works together to create the whole. Technology is much more than simply the equipment; it includes the meaning—or influenced understanding—experienced by both the artisan and his or her audience. Technology fulfills its purpose when shared with others—when it is in action. A piece of equipment whose switch is in the "off" position is simply an object; once powered on it becomes an available tool for narrating a story. Technology in biblical practice directs the congregation to participate in God's revelation and enables worship that is a proper response to Christ-revealed because of who he is and what he has done for them. Gayle Ermer poses the insight:

> If technology is a cultural activity that embodies human values, then what particular values should we be concerned about? . . . As in other areas of life, it is possible to begin with Christian values and arrive at quite different conclusions about how particular technologies should be designed and used. A Christian perspective calls for principled advancements and refinements that go beyond the mere technical considerations of efficiency and cost effectiveness.[23]

In this way, technology creates authentic worship when form meets function; technical artists demonstrate *tekhnê* when form and function align to present the gospel of Christ.

AV (Audiovisual) System. According to the Audiovisual and Integrated Experience Organization (AVIXA)—the official certifying body for the professional AV industry—an AV system is "two or more pieces of AV equipment designed to work together to meet a communication need. These systems can be connected with cable or wirelessly. The equipment used in the

---

[23] Ermer, "Responsible Engineering and Technology," 135.

system may be passive (not powered) or active (powered)."[24] The *CTS Exam Guide* continues: "In its simplest form, AV is about helping people communicate an idea effectively. . . . AV tools and technology are used to help people relate to and understand one another."[25] AV is the sound, light, and projection that assists in communicating and sharing ideas and experiences. Likewise, it includes atmospheric controls, ergonomics, safety conditions, theatrical effects, cameras, social media, the Internet, and other technologies. Today, collaboration and integrated environmental experience, along with virtual and augmented reality, are emerging trends within the audiovisual industry. For the church, an audio-mixing console, video switcher, projection, video cameras and IMAG, graphics, mic components, amplified sound, and LED or theatrical lighting design are the core components of the AV system.

Worship. The English word "worship" has a broad semantic range because it possesses both secular and religious connotations. In its simplest form, worship is the expression of admiration. In religious practice, it describes cultic practices in a systematic liturgical setting with expressions of thanksgiving, praise, communion, subjection, and petition on behalf of the worshipers toward the divine. Its etymology develops from the idea of "worth-ship," or according to Henry Coffin: "giving God His value, appreciating Him."[26] Larry Hurtado widens that definition to incorporate a "connection with reverence directed toward a figure that is treated as a deity."[27] Worship is elevating someone or something to the position of God or a god. In a negative way, for example, the golden calf was used by the Israelites as an object of worship in place of God, where YHWH demanded that only he is worthy of worship (Exod 20:3; Deut 5:7). This book holds a traditional view wherein worship refers to actions of reverence intended to express specific religious devotion to God within a known cultural context and religious tradition. It is the "actions by which people express and reaffirm their devotional stance toward [God] . . . while also affirming a positive relationship with the recipient."[28] In Christian circles, worship focuses on a way of life in praise of God, not simply a liturgical practice. Church technical artists are fundamental in creating a setting that promotes the congregation's Christ-focused worship. As will be demonstrated, for NT believers, Jesus is not only the object of worship but the person through whom and with whom worship is performed. Christians worship when they celebrate the life, death, and resurrection of Christ as the only one able to reconcile humanity to the Father as the propitiation for their sin. In worship, believers acknowledge their standing before God and their dependence upon him for holy living. The

---

[24] Grimes, *CTS Certified Technology Specialist Exam Guide,* 27.

[25] Ibid., 26.

[26] Coffin, "Worship," 218.

[27] Hurtado, "The Binitarian Shape of early Christian Worship," 188.

[28] Hurtado, "Worship, NT Christian," 910.

reading and study of Scripture, corporate singing, taking of communion, prayer, baptism, and fellowship, among other liturgies, serve as avenues within modern worship practice that the church technical artist is able to moderate technologically.

Worship Service: While each Christian denomination lends preference to its own traditions and practices over others, a "worship service" is customarily understood as the entirety of the corporate experience that may include the singing of songs, confession and thanksgiving, the Eucharist, preaching, offering of tithes, and baptism. Wayne Grudem's definition moves beyond glorifying God in voice, adoration, and praise to incorporate a deliberate act entering into God's presence: "The primary reason that God called us into the assembly of the church is that as a corporate assembly we might worship him."[29] The corporate response to the call to gather—not simply the church's designed practice—forms a worship service. Therefore, a worship service is defined as all gatherings of the church body, not only the "Sunday morning experience." This includes Bible studies, fellowships, accountability meetings, park days, church picnics, and other intentional collective activities. On a Sunday morning, a worship service is the complete experience of the congregants, from the moment they set their mind on attending "church," comprising not only the singing and preaching but also the parking experience, children's ministry, greeting and being ushered, donuts on the patio, and the technological encounter. In many churches the technological experience is not confined solely to the sanctuary but may include music on the patio, digital signage, and projected pre-roll graphic slides of upcoming events, setting the tone for the service. A worship service requires participation beyond attendance. Thus, a worship service is anytime the church purposefully gathers to come into God's presence in unity. In this way, a worship service is bi-directional; church leaders feed the body of Christ, and the body is to respond in praise of God while building up one another in fellowship.

SELECTION OF BIBLICAL TEXTS

This book surveys the current literature to determine in which ways the Bible might contain theological principles that can be applied to modern-day technical arts practice. Moreover, it asserts that practical worship motifs are ingrained throughout traditionally non-worship narratives in ways that modern scholars have either overlooked or simply not understood as possessing artistic context. When exposed, these findings can inform the modern understanding of worship practice, specifically technology in worship. A biblical

---

[29] Grudem, *Systematic Theology*, 1003.

scholar who is not an expert in worship arts would not naturally read the text through that lens, while a musicologist would most likely see the musical undertones in the text but miss its detailed theological substructures. This book bridges that gap in order to provide a vehicle for applying biblical principles to modern practice. This book does not seek to place the technical arts into the Bible—indeed the original authors could not have possibly perceived modern worship practice nor the utilization of electronic technology. It holds to Grant Osborne's concept of "story theology" wherein application to praxis is a holistic exploration of historical narrative, literary genre, contemporary contextual analysis, and which speaks to the foundations of the faith.[30] Therefore, the totality of the biblical texts had to meet certain criteria in order to extract concepts that could inform praxis:

View the Bible through multiple historical contexts: Old Testament, New Testament, ancient Near Eastern, the transitional early church, Jewish influences, and Greco-Roman influences.

Incorporate multiple genres and literary styles: The passages selected feature reference to narrative, history, epistle, pastoral message, directive, psalter, song, prophecy, hymns, poetry, imperatival instruction, commentary, musical score/notation, prayer, dialogue, thanksgiving, and petition.

Possess distinct artistic underpinnings according to biblical musicologists: If not fully developed, the texts contain at least introductory scholarly study in the context of contemporary worship practice, even if not commonly explored in that way by traditional biblical scholars.

With these understood parameters, three passages clearly separated themselves as worthy of exploration during the background literature review process.[31]

Exodus 35:30–36:1: During the documentary analysis of church technical artists, only two articles from the entire lot attempted to examine the technical arts theologically: "Exodus 31," by Mark Hanna,[32] and "Cultivating the Spiritual Side," by Mike Sessler.[33] Both of these articles explore the tabernacle narrative. Likewise, the concepts of work and creation were common topics in the study of worship arts. A number of texts presented ideas that drew parallels between the work performed by modern technical artists and the qualities held by the biblical characters Bezalel and Oholiab, even if not stated specifically as impacting the technical arts discipline. These include: Jeremy

---

[30] Osborne, *Hermeneutical Spiral*, 23, 28–29.

[31] It was not intentional to select only three passages. These three separated themselves among the all biblical references from the literature review of biblical worship and documentary analysis of the technical arts.

[32] Mark Hanna, "Exodus 31: Putting the Art in Tech Arts," *Church Production Magazine*, accessed June 22, 2015, http://www.churchproduction.com/story/main/exodus-31-putting-the-art-in-tech-arts.

[33] Mike Sessler, "Cultivating the Spiritual Side," Church Production Magazine, accessed April 15, 2015, https://www.churchproduction.com/education/cultivating_the_spiritual_side_of_being_a_church_tech_director/.

Kidwell (2016), Christ John Otto (2015), Patrick Sherry (2014), Richard Hess (2011), Tom Nelson (2011), Gene Veith (2011, 1991, 1983), Eugene Peterson (2010), Philip Graham Ryken (2006), and Leland Ryken (1989).

Hebrews 2:12–13: Two scholars explore the present work of Christ in relation to performance worship: John Paul Heil (2011) and Ron Man (2013, 2009, 2007, 2006), both doing so through an exegesis of the Hebrews text. Other scholars developed parallel arguments as Heil and Man in regards to Christ's role as the Mediating, Singing Savior, even if not to the same extent as they. Studies in this topic appeared to flourish circa 2011, yet faded, leaving multiple loose ends in the scholarship, specifically in terms of practical application. This phenomenon suggested that further exploration from the viewpoint of the technical arts could be well served. Key texts in addition to Heil and Man include: Frank Burch Brown (2014), Jeremy Begbie (2011, 2007, 1991), Steven Guthrie (2011), Michael O'Connor (2011), John Witvliet (2011), James Dunn (2010), Reggie Kidd (2005), Quentin Schultze (2005), and James Torrance (1970).

Colossians 3:16: While surveying worship arts practice—and particularly resources written by prominent Christian worship leaders—continuous reference was made to the phrase "psalms, hymns, and spiritual songs." However, it appeared that the verse was mostly cited in passing with very few references to the surrounding text. It is usually referenced out of context, not citing its position within the put-on put-off section of Paul's epistle. Likewise, it a popular "theme" to attach to practice in a majority of the worship works, even though not situated within a worship passage. It is rarely explored exegetically. James Janzen (2015) wrote a significant work, being the only text that exegeted the phrase from multiple vantage points, including its historical context, word meanings, and church practice, demonstrating the appropriateness of applying the text and the surrounding pericope to worship practice. This likewise suggested further exploration in terms of technical artists could be well served. Other fundamental resources that explore the theological-worship implications of the phrase include: Frank Burch Brown (2014, 2005), Barry Joslin (2013), David Toledo (2013), Jeremy Begbie (2011), Steven Guthrie (2011), D. A. Carson and Douglas Moo (2010, 2005), Ron Man (2007), Reggie Kidd (2005), D. A. Carson (2002), and David Detwiler (2001).

Examined together, these passages present fundamental characteristics pertaining to the people involved in the creation of the worship experience: they form a Christian theological understanding of church technical artists. Whereas passages could have been selected from the psalms or other more obvious "worship" verses, this theology of church technical arts contends the Bible says something more intrinsic about the nature of worship—and the people who create it—within foundational passages of Scripture pertaining to the faith.[34] All verses chosen are positioned within pericopae that contain greater theological significance in the scholarly discourse. By selecting these three passages, a fresh look at the biblical text, through the lens of both first-century Jewish and Greco-Roman worship practices—as theologians and musicologists today interpret such practices—will shine light upon the modern understanding of performance arts. It is not the belief of this book that biblical scholars to date are incorrect in their exegesis of these texts. Rather, it proposes these texts simply have more to say than previously articulated. They say something about "producing worship," offering theological validity to church technical arts in praxis.

---

[34] Only one other verse contained multiple references throughout the literature review: John 4:24, the concept of "worshiping in Spirit and Truth" now that God has left the physical temple (Ezek 10). This verse, however, has been extensively explored in both biblical studies and worship arts, and therefore will not be included in the same way that traditional worship texts were also excluded. Likewise, many citations of the John 4:24 passage were found in parallel reference to the other verses above and will be examined accordingly.

❧ 2 ❧

HOW DID WE GET HERE?
WHERE EXACTLY ARE WE?

INTRODUCTION

To establish how the technical arts fit into church practice, this book spring-boards off the current discussion within the realm of theology and the arts. Due to this being an artistic medium not yet formally explored in academic circles, the natural avenue to open the conversation on technical arts is from related foundational theological work already underway, by contextualizing how the technical arts relate to other art forms within the church arena, first historically and then specifically in the area already associated with music and aesthetics: contemporary worship leaders and songwriters.

In the twentieth century, North American fundamentalism used the new mediums of radio and television to shape the views of both the boomer and millennial generations. Idealistic debates over style and how church is supposed to be done lead to the largest obstacle the contemporary church had to overcome in achieving acceptance of today's new way of doing church. During this time of introducing the technical arts into the current arts, music, and theology conversations, a wide gap was created between the purists and the technological progressives. Even while modern society is exceedingly fascinated with the "latest and greatest"—with new technologies defining nearly every aspect of modern life—many churches are still balancing the pros and cons of integrating technology into their structured liturgical practices, either in fear of the service being watered down or the inability to integrate it appropriately. Marva Dawn suggests, "drastic changes in social fabric, caused by the onset of the technological milieu are intensified by the psychological reverberation of societal events."[1] These technological "tremors" were extensive, shaking norms of comfort. Advancements in technologies, music, and art have a record of challenging the church's views on worship ritual. Many in the church have been left wondering what to do with this new and exploding medium their congregations are interacting with daily outside the church walls. Do we embrace it or ban it? This question is not new; every new technology from musical instrumentation, artistic sculpture, stained glass, religious paintings, and architectural styles all faced this challenge before becoming the new norm.

Many "purists" are critical of modern church music, asserting that today's music and technological practices are inferior to classic hymns and hymnals of the previous century and that the church is dangerously close to losing an essential part of church history, due to the lack of theological content in today's songs. They claim traditional hymnody and psalm-based worship are to take priority in liturgical practices. Moving lights, loud thumping bass, and spoon-fed lyrics run dangerously close to transforming worship services into experiences of entertainment rather than reverence. Purists often call on the church to return to an older—and in their understanding—more "authentic" mode of practice. In a 2011 interview, John MacArthur presents this side:

> What you experience . . . here [at Grace Community Church] would have been exactly what you would have experienced if you had been here twenty years ago . . . or thirty years ago. . . . We pay absolutely no attention to the pop culture; we couldn't care less. We don't care what they're doing. It's irrelevant. We have a fixed point of reference, the Word of God. And I don't want to link arms with the culture. I want to link arms with the history of the church. I want to quote the great theologians. I want to sing the great hymns that generations of believers have sung and the reason we're still singing them is because they were so good. I want to link arms with the past. . . . It goes back several thousand years. I don't want people to think we just invented this.[2]

It could be asked, is the traditional format a biblical mandate or is it a personal preference of stylistic choice? MacArthur's negative outlook upon modern practice is surely not due to a false biblical understanding of worship,

---

[1] Dawn, *Reaching Out Without Dumbing Down*, 29.

[2] John MacArthur, "Practical Concern in the Local Church: An Interview with John MacArthur, GTY135," GTY.org, September 4, 2011, no pagination, https://www.gty.org/library/sermons-library/GTY135/practical-concerns-in-the-local-church-an-interview-with-john-macarthur. There seems to be a common thought among purists that what was considered church liturgy thirty years ago is inherently better than what is being performed now. That logic dictates that what was done thirty years before that is better than thirty years ago, and so on. Therefore, first-century practice is inherently better than any other worship practices. This is a failed logic because prior to recorded music it is impossible to fully comprehend worship practices and styles of previous generations. Though there is merit in understanding how the first-century church would have understood the biblical text in terms of worship practice, to make a qualitative better-than judgment is dangerous. Nevertheless, examining how worshipers at the time most likely viewed practice—and the characteristics of those who produce it—can inform contemporary technical artists regarding the qualities that ought to be brought into modern practice.

but rather possibly a constrained understanding influenced by a nostalgic belief in what church worship is and ought to be like, read back into the text two thousand years later. The question to ask regarding the use of technology in the worship service is not whether the church should mimic its history, but which history—that of ancient times or that of the modern church congregant? Today is also part of church history. So where *did* we start, and where *exactly* are we now? Throughout the history of the church, style has been closely associated with substance. If the main reason for the desire to recall practices of the past is due to the important adherence to solid biblical doctrine, then can it be that being biblical is and can be part of contemporary worship practice if style and substance are aligned with the biblical mandate?

There is no doubt the Bible places a special impetus on musicians and artisans to play at a high standard while performing in places of importance such as the temple and the throne of God. David wrote that his choirmasters were to produce a "joyful noise" to the Lord (Pss 95:1; 98:4; 100:1). The Chronicler described David's appointing of Levites as purposefully anointing them to perform "regularly before the ark of the covenant of God" (1 Chr 16:4–7). The NT cites the Apostles "singing hymns" as a method of evangelism (Acts 16:25), and the book of Revelation declares heaven as a place of genuine, submissive worship through harps and song (Rev 5:8–11; 14:2–3). Yet, throughout church history the relationship between the arts and the church has been storied, at times celebrated, at other times persecuted. The arts have played a significant historical role within both the ecclesial community and formal church services. Richard Viladesau notes, "a recurrent issue in the history of Christian reflection on the arts (even granted their theoretical legitimacy) has been the tension between art and asceticism."[3] At the risk of making the arts and technology serve modern pleasures, practitioners should consult both historical tradition and the concurrent cultural context. An educated view of the contemporary necessitates acknowledgment of the path that brought it here. It ought not be either/or, but both/and.[4] The historical perspective is indeed essential to contextualizing the current dialogue.

IN THE BEGINNING: BIBLICAL, ANCIENT NEAR EASTERN ERA

The earliest Egyptian findings on music cite it as "a form of spiritual communication (i.e., worshiping gods) before it became a pleasurable

---

[3] Viladesau, "Aesthetics and Religion," 33.

[4] There have been significant studies already completed in the realm of theology and the arts, particularly regarding the development of music and the arts in the church from a historical perspective by the likes of Jeremy Begbie, Frank Burch Brown, Steven R. Guthrie, Philip Stolzfus, Richard Viladesau, and Maeve Louise Heaney, to name a few. There is neither the need nor desire to add to their historical discussion as the scope of this research is concerned.

entertainment form."[5] The creation of music was purposeful rather than casual. Due to the close ties to the Israelites, this likely explains why OT examples of the use of instrumentation, skilled workmanship, and creativity in liturgical practice are many. Archaeological discoveries of literary texts, showing liturgical and ritual practices, suggest that Israelite worship was Canaanite in origin, which would have then had at least indirect influence from the Egyptians. The interaction between the ancient peoples can be viewed as developmental pillars for Israelite worship practices. Temple worship consisted of appointed musicians and singers following a fixed worship structure of playing, praying, singing, and sacrifice, similarly found in other ANE practices.[6] While early roots may have contained Canaanite and Egyptian influences, it was the Great Tradition ideal of cosmic order in music that dominated artistic ideologies throughout the region from Pythagoras to Plato, the early church, Augustine, Boethius, and well into the medieval era.

Pythagoras's (c. 570–495 BCE) discovery of the mathematical correlations between differing sounds created the belief that there must be some universality within audiological and musical compositions. Whereas modern society likens sounds in human terms of moods and emotions, the classical philosophers viewed sound within the hierarchy of natural order. Plato (c. 428–347 BCE) sharply criticized the emotional aspect of music, suggesting that it degrades the soul. Yet, at the same time he conceded that because music can also harmonize, it could assist in leading to a deeper connection with the universal by bringing disorder into order. By Augustine's time (354–430 CE), musical-mathematical Platonism became the prevailing belief. Universality in musical composition meant that skill and craftmanship within the process of creation would be prioritized over musical happenstance. While creating quality music was not to be strictly for pleasure, the belief held was that what is pleasurable is also aligned with its universal perfection. With the inverse also true: displeasing sounds could not possibly align with the universal.

It is commonly believed regarding early church practices that there was a decrease in the use of supplemental musical "equipment" in exchange for a predominantly controlled vocal worship style.[7] This change in the first and

---

[5] Williams and Banjo, "From Where We Stand," 197.

[6] From an evangelical worldview, God plays a special role in the development of worship practice. It can be dangerous to dismiss borrowed similarities. Israelite worship does not need to be either pagan or from God; it can be both. Pagan practices can be transformed for God's purposes and the edification of the church. A contemporary parallel would be the "set-up tear-down" church that gathers in high school gyms and community centers. Secular spaces are transformed for the worship of God for a limited period to integrate the secular in order to provoke the religious.

[7] This book proposes that this common understanding may be an incomplete picture of worship practice at that time. It argues that instrumentation was present in worship services, though modified to fit the context of the congregational loci. This fluctuates depending upon a pre- or post-temple-destruction context, and the physical location or distance from major pagan or Jewish centers. For example, church

second centuries would have emphasized home-based worship services, mirroring an informal synagogue style of worship that was entirely vocal and non-professional. In addition to the fear of persecution, which created internal tensions for societal conformity, David Detwiler notes the early Christians "lived in an environment of religious pluralism, [where] they coexisted with people who worshipped Anatolian, Persian, Greek, Roman, and Egyptian deities and with Jews who were devoted to the worship of one God and the observance of Torah."[8] This pluralistic setting created the need to be set apart from accepted societal customs in order to develop their own worship practices. The early church setting had two main purposes: the worship of Christ and the instruction of one another, with emphasis placed on Christ rather than the particular methods.

This transition to simplified and controlled worship practices came about for many of the same reasons contemporary worship wars were fought: finding identity. Jeremy Begbie suggests:

> Christians of the New Testament period do not seem to have had an antipathy toward instruments, but before long we see the stirrings of what was to become a vehement and sometimes extravagant polemic against instruments among the church fathers of the Western and Eastern churches, most of all because of associations with the music of idolatry and immorality in surrounding society—pagan worship, the theater, feasts, and brothels.[9]

The fear of idolatry troubled early church fathers like Augustine who believed that even though musical melody in vocal hymnody can be beautiful, the pleasure of its aesthetics could lure the worshiper's attention away from the truths it was intended to teach and ingrain. They believed that being caught up in the entertainment of the art of musical worship focused the attention on the art and artist, and away from Christ. This fear forced worship music to remain mostly monophonic vocal singing for the greater part of the next millennium. Yet, in favor of the arts in general, it is largely agreed that after the fall of the Roman Empire, the influence of the Catholic Church meant that the arts became spiritualized and redefined, pushing out common pagan ritualistic practice. The church seized the arts as its own. Artists who formerly focused on pagan rites became the originators of classic Christian icons and symbols. Skilled artisans became central storytellers of the gospel message for the next thousand years. What churchgoers could not understand in language, they could experience in art and architecture.

---

plants grounded in Greco-Roman culture would have more rituals borrowed from pagan festivals, whereas converted Jewish churches might incorporate extensive temple and synagogue practices.
[8] Detwiler, "Church Music and Colossians 3:16," 349.
[9] Begbie, *Resounding Truth*, 85.

REFORMING THE RENAISSANCE

Over the next thousand years there was little change in music and the arts until the High Renaissance of the late fifteenth and early sixteenth centuries when a flourishing of polyphonic symphonies, sculpture, stained glass, and classical architecture arose. The High Renaissance of the South and Reformation of the North dealt with the Roman Catholic Church through starkly different methodologies—the former addressed theology and the latter addressed art. Each complimented the another in order to serve the same ultimate purpose of bringing the worship of Christ directly to the people and away from the papal hierarchy. With the introduction of polyphonic singing, the Gregorian chant defined music for the longest period which, along with Latin hymnody, became the biggest influence of the early Reformers.

> Proceeding from Luther's basic understanding of music as a creation and gift of God, . . . the tradition he initiated drew on a huge range of material—including Gregorian chant, polyphony, sacred folk songs, and simple unison line singing—and led to an immense wealth of choral and instrumental music ranging from the work of Heinrich Schütz, through Johann Pachelbel and J. S. Bach, to major composers of the nineteenth and twentieth centuries.[10]

The Reformation provided a fresh look at the natural aesthetic world of music and the arts. Richard Viladesau points out an important change: "the strictures placed on sacred music by evangelical movements that stress God's word, like the Protestant Reformation, signify the attempt to assure that music serves as a vehicle for the word rather than constituting an end in itself."[11] Important questions that will define the growth of Western music began to be asked, as Jeremy Begbie suggests: "Is the created world being treated as able to glorify God in its own way, by virtue of its own distinctive patterns, rhythms, and movements?"[12] Can we treat music in its own right as a means of presenting theology? "Simply put: music is being grounded firmly in a universal God-given order, and thus it is seen as a means through which we are enabled to live more fully in the world that God has made and with the God who made it."[13]

Reformers like Martin Luther (1483–1546), John Calvin (1509–1564), and Huldrych Zwingli (1484–1531)—though not always agreeing on worship styles beyond vocal singing—opened the church door for the largest explosion in musical composition as scripturally based artistic works in themselves, only paralleled by the use of contemporary music and technology of the past decade. Stated positively, Luther appreciated the emotional aspect of art as

---

[10] Begbie, *Resounding Truth*, 105.
[11] Viladesau, *Theology and the Arts*, 31.
[12] Begbie, *Resounding Truth*, 92.
[13] Ibid., 94.

something distinctive, urging music to be an integral part of the church. Calvin took a sharp turn from Luther, desiring tradition, and objected to any instrumental accompaniment; all worship performance was to be limited to vocal singing. Even though he did eventually proclaim the importance of the Psalter, Calvin was predominantly apprehensive about anything that might compromise the otherness of God. Calvin explains:

> Care must always be taken that the song be neither light nor frivolous; but that it have weight and majesty (as St. Augustine says), and also, there is a great difference between music which one makes to entertain men at table and in their houses, and the Psalms which are sung in the Church in the presence of God and his angels.[14]

In other words, what we do during our congregational gatherings matters because they are performed to God and for the benefit of his people.

Vocal performance of the psalms was ultimately acceptable to Calvin because they are intrinsically biblical: "They are the words God gave us to praise him, and nothing can moderate music more effectively—nothing can better curb sin's power," notes Begbie.[15] Aware that worldly secular music would not be acceptable in the church setting, Zwingli—who ironically was an accomplished musician in his own right—suggested that music should be banned altogether from church worship. He rejected anything associated with the Roman Catholic Church; if the Catholics did it, he was against it. Zwingli believed that God had not commanded it; therefore, there was no need for it. Worship was to be private and internal in nature.

As we can see, the Reformation not only brought about major changes in the church proper, but in music, art, and the creation of worship. This was a time of questioning the strictures of the Catholic Church. Like seen today, the views of what was appropriate began to chasm.

ENLIGHTENMENT AND POST-ENLIGHTENMENT ERA

By the time of the Post-Enlightenment theologians of the nineteenth and twentieth centuries like Friedrich Schleiermacher (1768–1834), Karl Barth (1886–1968), Søren Kierkegaard (1813–1855), and Dietrich Bonhoeffer (1906–1945), polyphonic worship developed an overall acceptance. This period began to engage religious art through new mediums, flourishing in terms of stage works, concerts, interactive worship services, scientific exploration, and technology. Roger Lundin explains: "However much they varied in their approaches, these figures shared a deep desire to recover facets of the historic

---

[14] Calvin, *Preface to the Psalter*, no pagination.
[15] Begbie, *Resounding Truth*, 110.

faith that had been discarded or dismissed by the liberal tradition."[16] The arts were viewed as an acceptable medium with which to do so.

Schleiermacher viewed the arts as a religious experience: how you relate with art is how you are able to commune with God. Influenced by Romanticism, he "was at pains to make religion once again attractive to its 'cultured despisers,' . . . [locating] religion primarily in the realm of feeling . . . the 'feeling of absolute dependence.'"[17] Philip Stolzfus adds that "Schleiermacher's expression theory of music is no merely pedestrian language of the emotions. Music-making . . . is a mood-generating activity of the immediate self-consciousness, which manifests itself through public communication and performance"[18] where God is the direct expression of artistic ability and imagination.

In contrast to Schleiermacher, Barth viewed God as true beauty, an aesthetic that could only be for God alone. Voice and musicianship are to be used solely for God and the Christian community in connection with God in worship, celebrating the person of Christ. He firmly believed,

> Theology cannot be built on the belief in some supposed inner purity of human beings, nor on human inwardness; theology must rather focus attention rigorously on the way God has actually shown himself to be, which means a resolute attention to God's self-presentation in the person and work of Jesus Christ.[19]

Barth viewed his concept of God's word as the "Form of God's Beauty" in which theology is "a scientific examination of the trinitarian forms of revelation, proclamation, and Scripture [by which] the church becomes reified."[20] He believed the arts could serve a significant role in God's revelation. Barth viewed this no truer than in the works of Mozart, whose music he believed expressed theology better than the theologians of the day and that "for the true Christian is not mere entertainment, enjoyable, or edification, but [is] food and drink."[21]

Kierkegaard and Bonhoeffer both understood the difference between the arts as a call to action and the arts as expression. Kierkegaard did not deny the pleasure found in aesthetics yet warned about seeking it as a way to or substitution for salvation. Kevin Vanhoozer summarizes Kierkegaard's position:

> To listen in order to act, this is the highest thing of all. He goes on to compare worship to a theater and himself to a prompter who reminds those on stage of their lines. What the actor says matters because each

---

16 Lundin, "Doubt and Belief in Literature," 442.
17 Thiessen, "Artistic Imagination and Religious Faith," 82.
18 Stolzfus, *Theology as Performance*, 246.
19 Begbie, *Resounding Truth*, 153.
20 Stolzfus, *Theology as Performance*, 246.
21 Brown, "Musical Ways of Being Religious," 122.

word becomes true when embodied in him, true through him. People of faith who would speak understanding cannot therefore be content with speaking only.[22]

Kierkegaard believed that the biggest issues were not art per-se, but the aesthetic values wrapped up in the absolutisms of the Romantic movement of the time. The issue was not beauty, but the elicit feelings generated from created beauty that lead to focus on oneself and away from God. The issue as he saw it was not in the beautiful things themselves, but that when broken humanity experiences beauty, the response is to internalize it, filling oneself with pleasure, rather than give the glory to God in recognition.

Bonhoeffer understood all too well the struggles associated with being a follower of Christ. He served as a member of the underground resistance in Nazi Germany and ultimately was executed at Flossenbürg Concentration Camp for it. Jeremy Begbie and Steven Guthrie summarize: "[To Bonhoeffer,] music provides rich resources for discerning and articulating the nature of an authentically Christian engagement with a culture in which so much 'religion' had proven void and important."[23] As an accomplished pianist, Bonhoeffer saw a "polyphonic" connection between the arts and religion. Emotions weave together his almost fatalist, yet biblically informed, attitude. From jail he wrote that "it's a year now since I have heard a hymn sung. But it is strange how the music that we hear inwardly can almost surpass, if we really concentrate on it, what we hear physically."[24] In this way, the inner power of the arts to move a soul can be stronger than the physical impressions created. Yet, it is the conclusion of his May 9, 1944, letter to Renate and Eberhard Bethge that demonstrates best Bonhoeffer's assurance that theology and music go hand in hand. He closes the letter: "Are you going to choose the text for the baptism yourself? If you're still looking for one, what about II Tim. 2.1, or Prov. 23.26 or 4.18? . . . I'm sharing in your pleasure. Mind you have plenty of good music!"[25] Bonhoeffer trusts that liturgical practice necessitates "good music" alongside word-based theological delivery.

THE WORSHIP WARS BATTLES

When technology first entered the modern worship scene, the traditional notion of music as a ministry of the Word of God that developed out of the Post-Enlightenment deteriorated into an era known as the "worship wars." Contemporary worship leaders and pastoral leadership engaged in a lengthy

---

[22] Vanhoozer, *Faith Speaking Understanding*, 18.
[23] Begbie and Guthrie, "Introduction," 20–21.
[24] Bonhoeffer, *Letters and Papers from Prison*, 240.
[25] Ibid., 291.

debate over the proper distinction between what is true godly worship and what is simply entertainment. For the first time in over a century, common worship standards were challenged. The question had to be answered: How could the church engage and attract a changing millennial society without watering down the gospel message? The advent of multimedia production—which began as noticeably poor quality in production value—did not help the case of worship leaders and technical teams. This fight developed partially out of fear of leaving the known for the unknown, partially from the fear of losing the church's identity, and lastly from people on church technology teams lacking the proper knowledge of how to incorporate the new methods to the satisfaction of pastoral leadership. Because the church's identity is interweaved with its history, this new technological invasion—alongside changing worship styles—threatened that identity.

From the mid-1960s Calvary Chapel "Jesus Music" movement that allowed hippies and addicts to "come as you are," through to the Contemporary Christian Music (CCM) explosion of the 1990s that brought rock and pop music styles to mainstream airwaves as well as into the sanctuary, the line between liturgical tradition and relevance to present-day society was being blurred. Not until the modern hymn-based music compositions from the turn of the millennium that finally fused tradition with change did the differences held by both purists and progressive contemporaries begin to be resolved. Some believed the biblical idea that God alone is worthy of worship was being usurped by contemporary culture, even though the churches incorporating modern musical styles and technologies were still proponents of strong verse-by-verse exegetical preaching. Bryan Chapell recalls that during this transition from pushback to acceptance, every church he knew of was engaging in the worship wars to some degree or another. He cites twenty-two challenges facing the church at that time that compounded the problem beyond matters of musical taste:

> Transience of church populations, the demise of denominations, family breakdown, fewer people worshiping in the churches of their youth, aging church populations, concern to stem the exodus of a younger generation, the influence of pop culture, four decades of contemporary worship music, the charismatic renewal, ecumenism, technological innovations, globalization, megachurch influences, a longing for authenticity, the erosion of traditional values, reactive fear in much of the church, neoconservatism, fresh challenges to contribute to cultural transformation, a longing for anchors amid rapid cultural changes, rising interest in the global church, ancient-future church movements, and neo-Catholic movements.[26]

The world was changing and the church was stuck between assimilation

---

[26] Chapell, *Christ-Centered Worship*, 125–26.

and rebellion.

How could the church reconcile being set apart as God's people from a society that was moving away from its core traditional values, while at the same time being all things to all people (1 Cor 9:19–23)? Some churches, like Calvary Chapel Costa Mesa in Santa Ana, CA, Willow Creek Community Church in South Barrington, IL, and Saddleback Valley Community Church in Lake Forest, CA, discovered the answer: view technology and music as a gift from God that assists in presenting the gospel message, and not a device of a corrupted society. Steve Miller suggests that those "who scoffed at Jesus Music . . . brushing it off as a passing fad should take a second look at music that has become a major vehicle of both evangelism and discipleship."[27] Similarly, Willow Creek and Saddleback Church employed "seeker services" that welcomed congregants to worship through musical styles they understood and listened to the other 166 hours per week that they were not sitting in the Sunday morning sanctuary. The "seeker" strategy had a specific goal in mind:

> Each [song] has a purpose. Music is chosen for its ability to touch people, to make them think, ponder, question, or experience the truth. The staff know which radio stations the unchurched listen to and what concerts they attend, and use the styles that move [them].[28]

Even though successful in engaging the unchurched, these new movements experienced pushback from the traditional church. During the height of the worship wars, the belief was that if musical styles and technological production techniques were allowed to mirror society in any form—notwithstanding the gospel message buried within—idolatry of the world and culture would be presented over the gospel itself. In other words, they believed turning away from tradition meant turning toward idolatry. Bob Kauflin of Sovereign Grace Music responds to these critiques by noting that even though we have been in "conflicts over musical styles, song selections, and drums . . . far too little has been said about the worship wars going on inside us. And they're much more significant. Each of us has a battle raging within us over what we love most—God or something else."[29] The missing element was sincere communication and understanding that worshiping God, not musical styles or technology, was the common goal for both sides of the argument. Each could have learned from the other. In the process of gaining cultural acceptance, the traditions of the church were turned aside, perpetuating a cycle of cynicism, first through separation, then flailing integration, and finally conflict. "Both sides are equally simplistic," concludes Tim Keller. "Contemporary worship advocates consult the Bible and contemporary culture, while historical worship advocates consult the Bible and historical

---

[27] Miller, *The Contemporary Christian Music Debate*, 177–79
[28] Miller, *Worship Leaders*, 2534–36.
[29] Kauflin, *Worship Matters*, 21.

"[30] Even with agreement on biblical centrality, this lack of commu- between the traditionalists and contemporaries veered down a path of quarreling over style, dividing the church body, rather than recognizing the clear agreement on the gospel message as the core ingredient to each.

When the "war" was won, it was by neither the contemporaries nor the traditionalists; it was a synchronous blending of the contemporary incorporating the traditional. Neither side lost the worship war; each retained a voice. The precautions on both sides of the aisle helped to develop the essential aspects for today's worship leaders, songwriters, and technical artists. Facing music and technology with clean hands, a heart of worship, love, and understanding of who God is and what he has done are essential to the modern church service: concert lighting and all. According to Tim Keller,

> The solution to the problem of the worship wars is neither to reject nor to enshrine historic tradition but to forge new forms of corporate worship that take seriously both our histories and contemporary realities, all within a framework of biblical theology.[31]

The overarching acceptance of contemporary music—and simultaneously technological advances—was the product of a changing focus by worship leaders themselves. As ministers of the Bible, incorporating scripturally rich and theologically sound lyrics became a priority alongside developing well-trained musicians and support crew. The result: modern hymnody. Almost ironically, Mikie Roberts cites Grove Music's generic hymn definition as "unknown [in] origin but first used in ancient Greece and Rome to designate a poem in honour of a god. . . . In the early Christian period, the word was often, though not always, used to refer to praises sung to God,"[32] just as it became post-worship wars.

Matthew Sigler's examination of CCLI's "Top 25" list over the past few decades demonstrates the change in musical style which led to the current musical and technological preference. He finds,

> The majority of the songs [on the initial 1997 list] are published or administered by Maranatha! Music, founded in 1971, by Chuck Smith. The number one song on this first Top 25 list, "Lord, I Lift Your Name on High," is also the newest, published by Maranatha! in 1989. [It] would remain [number one] . . . until August 2003 and become what was for many the quintessential contemporary worship song.[33]

This finding suggests that the first wave of CCM songs incorporated into the church became the new "traditional" for the present-day church. The same music that was being criticized twenty-five years earlier had now become the expectation. The songs that founded the new genre were indeed

---

[30] Keller, "Reformed Worship in the Global City," 197.
[31] Ibid., 198.
[32] Roberts, *Hymnody and Identity*, 4.
[33] Sigler, "Not Your Mother's Contemporary Worship," 450.

not all that new. Subsequently, in February 2005, a hymn arrangement "The Wondrous Cross" appeared in the Top 25 for the first time; it was a contemporary arrangement of "When I Survey the Wondrous Cross" by Isaac Watts, the eighteenth-century English hymn writer. Songwriters like Matt Redman, Bob Kauflin, Paul Baloche, and Darlene Zschech began to break through the new songwriting standards by being able to capture the essentials of the Christian faith within the context of corporate hymn-praise based worship. The songs written had a direct impact on the teachings of the church. Out of this spawned the "worship pastor," a musician with a role equally balanced with leading corporate worship and shepherding the congregation from the performance platform. By incorporating the gospel message of salvation and faith in Christ into the vast musical genres common to contemporary culture, like pop, rock, country, gospel, rap, and punk, among others, Christian music has taken on the same properties of secular music in recognition that one size does not fit all when it comes to connecting to and worshiping God. As a result, the number of Christian radio stations has more than doubled each of the last two decades, "making CCM radio the fourth most common radio format in the United States and the second most common music format,"[34] as well as the fastest-growing segment of the music industry.

This blend of modern music styles with reflective hymns and biblical truths won the worship wars. "Whereas the early years of contemporary worship had often been defined by a suspicion of tradition, including hymnody, there was instead in the new millennium a rethinking of this radical rejection of the past,"[35] which is now integrating itself into modern culture in order to reach the next generation. Music and technology in the church have always reflected the immediate cultural bias, whether or not the church itself was a willing contender in incorporating it. The "wars" over worship are usually over style and not about the essence of worship itself. Popular worship songwriter Paul Baloche summarizes the battle, writing that regardless of the worship style, the arts ought to "point us to the point of it all—pointing us to the Savior."[36]

MODERN CULTURE AND THE TECHNICAL ARTS

Just as contemporary believers view beautiful sunsets and natural wonders, declaring God's majesty behind creation (Ps 19:1), non-believers too may gaze at the heavens, recognizing there must be something bigger than self at work. When a person allows himself or herself to be captivated by the

[34] Bentley, "Uses and Gratification Study," 3.
[35] Sigler, "Not Your Mother's Contemporary Worship," 453.
[36] Kauflin, *Worship Matters*, 11.

majesty of skillful high art like the Mona Lisa or the magnificence of a Mozart symphony, they can be swept away to a new reality, an out-of-body experience of sorts. These certainly are "spiritual" experiences, even if not wholly religious in nature. Therefore, when believers allow beauty in both Creation and their creative acts to inform their craft, the door opens for soul transformation and sanctification. Already, several mainstream worship leaders are embracing this opinion, constructing set-lists based on a song's biblical integrity as much as musicality. In this same way, Tim Keller had welcomed non-Christian, professional, paid musicians to perform in his church worship services in order to provide the highest quality for his congregation, just as they would enjoy at a Broadway production down the street from his Manhattan, NY, based church. "We have made it a basic principle that music in corporate worship must be of high technical and artistic quality, as well as theologically sound and fitting."[37] These priorities offer congregations a worship experience that encourages them to fulfill their liturgical roles as active participants. This philosophy, though possibly agreed to in academic and theoretical discussion, is far from being incorporated into daily practice.

Modern Christians do not live in an artistic vacuum. They are continuously inundated with sensory experiences through the workplace, television and radio, personal computers, and social media. Likewise, Quentin Schultze adds, "Just as electronic machines impact the world, so do our sentences, artworks, novels, and films. . . . Visual and verbal symbols, not just words, can alter our worship for good and bad."[38] Thus, this demands a higher level of responsibility when engaging the audience from within the church context so that what is presented is unquestionably doctrinally sound.

In the same way, virtuoso cellist Yo-Yo Ma suggests that "feeling and emotions as expressed in art and music play a central role in high-level cognitive reasoning. . . . These discoveries suggest that a new way of thinking is possible,"[39] one that allows the arts to foster true spiritual growth and a "new enlightenment."[40] The arts for Christians could be used not only for joy, healing, and cognition but a greater restoration, that is, biblically informed sanctification, with technical artists serving integral roles in the church's spiritual development as the intermediaries between stage and congregation. In the act of "producing worship," technical artists must be keenly aware of the theology they produce. For example, lowering the air conditioning and sweeping lights across the congregation can produce the "feeling" of the Holy Spirit "touching" the congregant by creating goosebumps and

---

37 Tim Keller, "Reformed Worship," 236.

38 Schultze, *High-Tech Worship*, 42–43.

39 These thoughts are backed by recent scientific studies that have demonstrated that musical and artistic response are a few of the only activities that stimulate the entire brain simultaneously. See Alluri, et al., "Large-Scale Brain Networks Emerge," 3677–89.

40 Ma, "Behind the Cello," no pagination.

metaphorical blinding white lights, yet technical artists must not confuse feeling with true transformation. Engaging people means changing people, and specifically changing how they interact with God during the worship experience.

In this way, the case can be made that God as Creator endowed humans as *imago Dei* with the capacity for creativity and gave the arts as a tool to glorify and worship him while leading other worshipers to do the same. One cannot be learning about God without learning from God.[41] Thus, utilizing creative skills endowed by God can bring worshipers into further knowledge of him. In light of this, churches may be well served to further explore all available creative options, including technological ones. Surely God has used popular styles of the day to further his plan, church, and people; should not then, technology—and creative technical arts—be accepted into church liturgical practices? "Style" is traditionally less about taste than it is about appropriateness. The style of an artistic work was to align with societal norms, the context with church doctrine. In this way, when the technical arts are used in church practice they too possess the ability to create unity and harmony in the church exactly through employing the technological methodologies utilized by the congregation in their daily practices. The method becomes the message. It can, therefore, be argued that scholars and practitioners ought to approach studying all methods in which theology and the technical arts can be appropriately applied to worship practice, for as Philip Stolzfus notes, the arts are not as much "an individual experience of feeling, but as a cultural performance of group solidarity through the critically appropriated and participatory constructed phenomenon of attunement."[42] The theological and the aesthetic are brought into harmony for the benefit of the group. Utilized properly, then, the technical arts can become a medium for believers to be further conformed into the image of Christ. Because medium can be the message, the same ought to be true of the medium of technical arts. Since the technical arts are a vehicle for delivering the gospel, they are able to be seen and studied in that light.

Technical production is an art that is created as it is performed. It is temporal. Therefore, it is imperative technical artists understand what is actually being created and recreated and how their craft informs the greater theological school of ecclesiology. Thus, it could be suggested that—like with traditional church artisans—for technical artists to be effective they must recognize the baseline biblical truths that their art is supposed to represent, both practically and theologically. Daily habits inform how church technical artists approach their craft.

In what way, then, do the main current practices utilized by church

---

[41] Begbie, *Resounding Truth*, 20.
[42] Stolzfus, *Theology as Performance*, 251.

technical artists inform the theological understanding of the craft? If church techs are often working during the sermon and ministry times, how much focus can be specifically put toward comprehending and integrating the message from the pulpit? If technical artists are following and communicating with each other on social media during the sermon messages and work hours, can they be gleaning as much from the message as other congregants? One church tech responded during LeadLab San Diego 2015 that he may not pay attention to everything, but after he has heard the sermon three times over the weekend, he "got it all." This demonstrates that how church technical artists learn theology and grow in the knowledge of their faith is fundamentally different than traditional church attendees. Likewise, corporate sermons do not generally focus on the biblical principles of church technical arts, but rather on Christian living. Rarely, if ever, is focus placed on direct application to the technical craft from an exegetical perspective. Nevertheless, Christians would hold to 2 Timothy 3:16–17 that all Scripture from God would be useful in equipping them for their work. Thus, all learning that grows one's knowledge of the faith would in some way impact how technical artists approached their craft, even if the biblical knowledge is not directly related to the individual actions at hand, like operating the audio console or changing camera shots. Nevertheless, this theology of church technical arts argues that what church technical artists do in practice is more than simply running a church service, but also directs the church's ecclesiology. They have a tangible role in the church's future, and remain an underserved demographic within the church itself.

TECHNICAL ARTS AS A THEOLOGICAL DISCIPLINE

The scholarly conversation regarding the worship aspect of theology and the arts is still relatively new, only gaining traction over the past few decades. As a general topic, it is flourishing among a handful of colleges and seminaries like the Calvin Institute of Christian Worship at Calvin College (Grand Rapids, MI), Dallas Theological Seminary (Dallas, TX), Duke Divinity School at Duke University (Durham, NC), Fuller Seminary (Pasadena, CA), Harvard Divinity Schoool (Cambridge, MA), Southwestern Baptist Theological Seminary (Fort Worth, TX), University of Birmingham (Birmingham, UK), University of St. Andrews (St. Andrews, UK), and the Yale Divinity School at Yale University (New Haven, CT). Nevertheless, theological training specifically for technical production is lacking. This is likely not purposeful; it is simply not where the interest and conversations have been directed. Frank Burch Brown confirms this finding:

Religion and the arts . . . has not often been given much visibility on the map of religious studies, and has never been charted in detail. This means that relatively few readers are likely to be an expert in a given area. . . . [Omissions] may also be able to be addressed in the future— such as those that are the unintended result of editorial blind spots or *those that involve digital media only now coming more fully to the attention of scholars.*[43]

It is notable that Brown specifically notes the lack of study of digital media, an area that this book intends to fill.

Even though a significant amount of academic material on "worship" and "religious aesthetics" does offer passing references to technological practices, the specific focus on technology and the technical artist is surprisingly absent. Not since Quentin Schultze's foundational book, *High-Tech Worship?: Using Presentational Technologies Wisely* (2004), has there been any substantial work explicitly geared toward technical artists that theologically examines the integration of technology into worship. A few key tutorials like Len Wilson and Jason Moore's *The Wired Church 2.0* (2008), Brad Herring's *Sound, Lighting and Video: A Resource for Worship* (2009), and Gregory Zschomler's *Lights, Camera, Worship!* (2005/2014), offer legitimacy to the practice of integrating the technical arts into structured church services through presenting "best practices" for church technical artists to follow. However, they serve little purpose in contributing to the scholarly discussion. The closest attempt at a biblical understanding of the technical arts is Lee Bloch's *Worship from Backstage: A Biblical Perspective* (2008). Bloch states that his goal is "to be a battle cry, a wake-up call . . . [to] take a stronger stand for Christ . . . [and] reveal some practical and biblical methods you can employ." The attempt is encouraging and demonstrates a desire to align the technical artist with his or her faith, but it is not a "biblical theology" of the technical arts, but rather a personal expression of belief.[44] It is, nonetheless, a solid step forward.

Two practical resources are worth noting. First is the recent development of "MxU Coaching," an online small group devotional Bible study ran by Andrew Stone, Jeff Sandstrom, and Lee Fields. It is the first mainstream attempt by prominent professionals in the church tech industry to bring together improvement of craft with a biblical understanding of it. It incorporates a weekly devotional, a small group deep dive into "what the Bible has to say about what, and why, we do what we do as we lead our teams,"[45] a monthly video conference blending spiritual application with topical discussion, and various office areas and chat lines for connecting, praying, and question answering. Second, every Monday morning, *Church Production Magazine*

---

[43] Brown, "Introduction," 4–5, emphasis mine.

[44] Bloch, *Worship from Backstage*, 134–39. In a verse-point structure rather than exegesis that allows the text to define the theology, Bloch explores popular complaints and trials that technical artist experience and offers solstice in the form of a topical biblical perspective. For example, in the section "The Call," Bloch gives a personal reflection rather than an exegesis of biblical narratives on "calling," of which there are many. The work presents statements of belief rather than exploration of biblical text. Bloch reads the craft into the Bible rather than allowing the Bible to inform the practice. Even though most evangelicals would likely view the book as "biblically correct" and aligned with the tenets of the faith, it is a personal expression of belief and not built upon textual exegesis.

[45] MxU Rocks, "Coaching – MxU," no pagination.

releases a weekly devotional email newsletter with articles written by prominent tech leaders.[46] These offer insight to common struggles tech leaders face from a biblical perspective. They offer encouragement, understanding, and tips to better lead tech teams biblically.

Now over a decade removed from Schultze's work, theologically-specific resources on church production are still lacking, yet the prominence of technology in people's daily lives is making them both expect technology in worship, as well as the high-quality use of it. Popular worship leader Stephen Miller calls technology "a wonderful servant, but a terrible master."[47] Even so, the current demographic change suggests it is both a societal and sanctuary given, inescapable for living, thriving, and capturing the "selfie" generation. Given its importance, the lack of theological works on this subject is startling. Though still relatively new in development, two United States megachurches have expanded their worship training departments to include a certification in the technical arts, with Gateway Church's School of Tech Arts (Southlake, TX) and Calvary Chapel Costa Mesa's School of Worship Media Arts (Santa Ana, CA). Likewise, a number of Christian colleges like Point Loma Nazarene University in San Diego, CA, California Baptist University in Riverside, CA, Azusa Pacific University, in Azusa, CA, Anderson University in Anderson, IN, Greenville University, in Greenville, IL, and Messiah College in Mechanicsburg, PA, are adding commercial music degree programs integrated with their worship arts programs.[48] It is still to be seen how these churches and universities will impact the greater technical arts scene, but it does appear promising. Nevertheless, with the arrival of such programs, opportunities do exist for technical artists to enter the scholarly conversation.

CONCLUSION

Observing the changing views in regards to music and the arts over time, the constant struggle between society and the church becomes the defining factor. The church's view of culture forms its opinion of structured liturgical methods. As views have changed throughout church history, so too has application. To this end, the question needs to be answered: as proponents of historical worship, whose and which history? All "traditions" were at one time or another new themselves. The arts and the church have worn many masks throughout time in a continuous struggle of balancing proper liturgical practice and cultural influence. Viewing technical arts from this historical

---

[46] In full disclosure, I am a contributor to the *Church Production Magazine* weekly devotional.
[47] Miller, *Worship Leaders*, 110.
[48] This list is not intended to be a comprehensive list, but rather to demonstrate that the combination of tech and worship in higher education is becoming a common trend.

perspective, the question is not *if* styles that mirror contemporary society should be incorporated, but *how*? Indeed, history has shown that the church has most often adapted to mirror culturally stylistic norms. Laurel Gasque ties such adaptation to the current discussion, writing: "theologians [of the past] could differ over doctrinal positions, but they all practiced the pieties induced by art,"[49] regardless of the selected—or not selected—style. "That art provided a unifying, rather than a divisive experience in the ancient church is a far cry from the worship wars of today that divide congregations over musical styles!"[50] As artistic theologians throughout history have demonstrated, integrating the current artistic cultural milieu into their normal worship practices can unify the church body to engage in a proven approach to spreading the gospel message. Today, that medium is the technical arts.

---

[49] Gasque, "The Christian Stake in the Arts," no pagination.
[50] Ibid.

ॐ 3 ॐ

EXODUS 35:30–36:1: BEZALEL
AN OT TECHNICAL ARTS MODEL

35:30 Then Moses said

    to the people—the sons and children—of Israel,
   "Behold! The Lord has specifically called
    by name Bezalel

      (son of Uri, grandson of Hur, of the tribe of Judah),
35:31 and he has filled him

    with the Spirit of God, in regards to:
     skill and wisdom at heart,
     intelligence and understanding,
     knowledge and expertise,
     and all craftsmanship,
35:32 to devise artistic designs for masterful work...[1]
35:33   in every skilled craft

      (by any sort of workman or skilled artistic designer).
35:34 And the Lord has inspired he and Oholiab

     (son of Ahisamach, of the tribe of Dan)
   to teach their skills to others.

---

[1] Some reference to the materials and objects made by Bezalel, Oholiab, and the craftsmen, will be noted. The actual artifacts are less important, and can apply to all craftsmanship, even today. Yet, it should not be lost that the items made are for the dwelling place of God and for the proper worship of YHWH. The items are specifically chosen to be royal and holy. For the sake of this book, the characteristics of the people—the artisans themselves—and not the items of the Tabernacle, which already has a longstanding history of scholarship, is explored. See Strong and Sarna for a complete study of the tabernacle and its contents.

35:35   He has filled them with skill to do every kind of work...
                and given ability and skill
                        in the hearts of all able and gifted artisans,
                so that they may make all that I have commanded.
36:1   Bezalel and Oholiab and every skilled craftsman
                in whom the Lord has personally put skill,
                        wisdom, intelligence, and understanding,
                to know how to perform
                        any and all work in the construction of the sanctuary—
                        and all that is Holy—
                shall work in accordance
                        with all that the Lord has commanded."

INTRODUCTION

Twenty-six times in Exodus 25–40, the section describing the construction of the tabernacle, the word "skilled" is used to refer to the way something is made or a characteristic it possesses. The narrative of Exodus 35 (cf. Exod 31), where the tabernacle is commissioned for construction, specifically paints a picture of the relationship between the craftsman and the church as well as between the craftsman and God. Through the description of the biblical characters Bezalel and Oholiab, the text presents a portrait of a biblical artist and the six distinct attributes he or she is to possess: Spirit, skill, intelligence, knowledge, craftsmanship, and teaching ability. In Exodus 31 God informs Moses at Mt. Sinai that he has personally selected Bezalel to construct the tabernacle. In chapter 35 Bezalel responds to the call. Bezalel is the master builder and chief artisan of the tabernacle, its furnishings, and the clothing; Oholiab serves as his anointed assistant. Eugene Peterson comments regarding Moses and Bezalel's relationship in the final chapters of Exodus: "Bezalel is in charge. And what he is in charge of is making provisions for worship. . . . At chapter 35, Moses steps aside and hands things over. . . . Bezalel provides the people with the material means for worshiping through the wilderness and living in the promised land."[2] The Lord gave the instructions to Moses; Bezalel responded to an internal calling. Moses fully recuses himself, so that Bezalel can perform his inspired tasks.

    The purpose of the tabernacle is for God and humanity to commune with one another. Not since the garden does God create for himself a physical place to dwell with his people. It is the earliest record we have of something

---

[2] Peterson, "The Pastor," 94.

constructed for the worship of God, with instructions that come from God. The general idea was that God had become a wanderer with his wanderers; he became the "tent" for his fellow tent-dwellers. He came down to humanity so they may have fellowship with him. The tabernacle served as the portable home for YHWH until Solomon built the temple on Mount Moriah approximately two hundred years later.

The tabernacle narrative is split into four parts: (1) the list of building materials; (2) identification of the builders; (3) the construction and making of furnishings; and (4) the census and tax levy to support the tabernacle's cultic functions. If God desired, he could have either built the tabernacle himself or made anything anywhere to be the meeting and worship place of God. Instead, he chose inspired artists to create his dwelling space while equipping them with the necessary skills, along with the Spirit, for a task that would meet his holy standards. In this way, the arts received divine approval. To an untrained reader, the passage contains endlessly monotonous details, yet it is essential to recognize that an architect could not effectively build the tabernacle based on the description given; the proper details are not present.[3] God gave the instructions to build the tabernacle—not instructions for the tabernacle—and bestowed upon the artisans the ability to do so through the inspiration of the Holy Spirit.

Bezalel—the first person in the Bible said to be "filled with the spirit" (Exod 35:31; 31:2)—is an often-overlooked character in the biblical narrative. Modern commentaries often skip or summarize the ten chapters pertaining to the construction-commissioning account, while focusing on the earlier works of Moses. Yet, Bezalel is a significant character for NT Christians because of the parallels between him and Christ. The tabernacle is a central space for Israelite worship wherein Bezalel's actions play a significant role in how God's people worship and experience God. Bezalel produced Israelite worship through an artistic creation of divine inspiration. This chapter explores the tabernacle construction narrative specifically focusing on the person of Bezalel and his assistants, Oholiab and the tabernacle artisans. For contemporary church technical artists, these biblical characters offer a template for the qualities necessary to perform their craft within the confines of God's sacred space.

EXAMINATION OF THE IMMEDIATE CONTEXT

The entirety of Exodus 25 to Numbers 36 is about organizing the proper worship of a holy God. The tabernacle construction and commissioning

---

[3] Kidwell, *Theology of Craft*, 25; Sarna, *Exploring Exodus*, 191.

narrative lies at the center of this third high point in the book of Exodus, which has more time devoted in Scripture to it than any other object, making up all or a majority of eleven chapters, 25:1–31:18 and 35:1–40:38. Chapter 35—Moses's informing the people of Bezalel's commissioning of YHWH's dwelling place in the midst of his people—is the fulfillment of 31:1–11, God's informing Moses of his plans. The duality behind the two tabernacle descriptions shows that God presents the meticulous details and then that the directives were to be carefully followed and carried out.[4] The tabernacle was not simply a structure for worship, but a systematic ordering of worship experience: The sanctuary—"the holy place"—sat at the center of camp, placing YHWH at the center of their attention. The courtyard contained the bronze altar to the east, with the two rooms of the tabernacle to west end, "the holy place" and "the holy of holies." In "the holy place" sat the altar of incense, the table of showbread, and the menorah, while in "the holy of holies" lived the highly adorned ark of the covenant, as well as the enthronement of YHWH over the ark. Each aspect plays a role in renewing God's covenant with his people. Meeting and dwelling express a place where God and humans can find reconciliation.

There is an oft-overlooked characteristic at play: the Israelites were not only able to flee from the rule of Pharaoh, but were simultaneously able to inspire great works of art and design during a time of change and doubt. On the first anniversary of the deliverance from Egypt, the tabernacle is consecrated. Because Moses neither understood nor possessed the ability to transcribe the blueprints into being, God filled Bezalel and Oholiab with the Spirit for the work of building and creating within the parameters of the Second Commandment not to create any carved image of heaven or below that would be worshiped in place of him. God gave Moses the directions, but he was not the one to perform the task.[5] He was a prophet, not an artist. By setting the delivery of the tablets directly prior to the tabernacle narratives, the text places Yahweh in a position of leadership through the entire construction event, not Moses. Understanding the text in this way makes sense because Moses is not an artist; God would have to direct and select his chosen

---

[4] Victor Hamilton (*Exodus*, 409) suggests there are narrative similarities between Moses and Joshua, Bezalel and Oholiab, David and Solomon, and Solomon and Huram-abi. In each case, the first person planted and the second watered. There is a continuation of fulfillment and success. "The analogy [is that of] a skilled master builder [who] lays the foundation, and then someone else builds on it."

[5] Mark George (*Israel's Tabernacle*, 164) suggests that the tablets given to Moses in Exod 31:18 were not merely the Ten Commandments, but the whole conversation between Moses and God, which would have included the directions to be handed over to Bezalel for the construction of the tabernacle. He suggests this idea is confirmed by 24:4's citation that "all the words" were written down. I find this to be a leap but noteworthy nonetheless. It is a view that does demand some consideration. If it were the case it would add further implications for both God's command to build and the methods that Bezalel and Oholiab would employ in the process of construction. Likewise, George argues that the word used for "covenant" preceding the construction narrative is never used until later in Exodus 35 when action is taken to build YHWH's dwelling place.

builders. The descriptions are so exact—yet at the same time so abstract—that it would take a skilled person to understand how they come together. That is a job more suited to God—and his representatives Bezalel and Oholiab—rather than to Moses. The tabernacle needed the most qualified artisans and finest materials. The building of the tabernacle means that Bezalel is not only responding to God's spoken Word; he is also authorizing performance of artistic worship within in it.[6] He is creating a new realm of reality for God on earth that is intended to not only be viewed, but experienced. In this way, artistic creations formed within sacred space, possess an intrinsic performance value useful in fulfilling their intended purpose.

In contrast to the fulfillment of God's physical dwelling within the midst of his people, the golden calf narrative is purposefully placed between the two commissioning narratives to emphasize that not all creations are sacred and poetic. The religious leaders decide to create an object of worship, but God desired to become the object himself. In the narrative, through the commissioning, rebellion, and construction, Israel experiences both God's anger and his mercy. The anger of rebellion transforms to mercy through commissioning Bezalel, the technical artist, with the command to erect God's physical presence among his people. While Aaron created a representation of God in the form of the golden calf, Yahweh was expressing his desire to become the embodiment of meaningful worship for the people. Aaron and the Israelites did not need to create an object of worship but rather a sacred space for God to be that object among them. Demonstrating repentance, Bezalel, Oholiab, and the people of Israel, showed full obedience in following God's ordering of the tabernacle. Both Moses and God express their approval of Bezalel and Oholiab's obedience. Moses does through blessing the artisans (Exod 39:43); God does by filling the dwelling place with his glory.

THE TABERNACLE: FACT OR FICTION

Any discussion of the construction of the tabernacle must include an examination of the scholarly debate on whether the tabernacle actually existed or is merely an act of literary fiction. Whether it did or did not does not impact the theological value for Christians, who would view the narrative as inspired and suitable for teaching (2 Tim 3:16). Likewise, the outcome has no direct effect on this argument or validity of what the Bible asserts about

---

[6] Peter Leithart ("Making and Mis-Making," 316) cites Moses as the one who built the tabernacle. This is a common theme among the commentaries. Bezalel and Oholiab are often overlooked as the builders in favor of a continuing narrative of Moses's authority over the project. Even though Moses may have commissioned the project and been the one to "sign-off" on its completion, the construction itself was reserved for the technical artists: Bezalel, Oholiab, and their assistants.

technical artistry. Yet, what it says about artistic practice has the possibility of informing modern praxis. Therefore, following is a short explaining the main arguments both for and against the physical existence of the tabernacle.

The main argument for the existence of the tabernacle lies in the fact that "holy tents" were a common ancient Near Eastern practice. In fact, the description given could be considered quite typical and mirrors other ANE accounts of temple-building and worship shrines from neighboring peoples. The tabernacle itself mimics other known worship spaces. The words *mishkan* and *'ohel* point to nomadic origins that can be dated back to Israel's settlement in Canaan. Parallel uses of "tent dwellings" exist in the Ugaritic, Akkadian, and Targumic Aramaic languages. Archaeological evidence throughout the ancient Near East, including Ruwala tribes, Islamic Near East, Bedouin tribes, Phoenicia, Syria, the Midianites at Timna, and Ramses II of Egypt, which included an exact replica set up as his "war tent," show analogous descriptions to that of the tabernacle. The use of jewels and ornamentation for decoration was popular in ANE cultures.[7] For example, items that are integrated into both the tabernacle and El's tent include: appellations and wooden supports, multiple large rooms, the fine fittings of gold and silver, throne, footstool, libation utensils, table, construction by a fine craftsman (Kothar in El's case), and accounts of rebellion similar to the golden calf incident that interrupted the work. The typical ANE pattern of tent and temple construction narratives include: (1) divine reason for the construction with the gods' consent; (2) direct transmission of the command to build to the people in charge of construction; (3) preparation by the acquiring of supplies; (4) the construction account; (5) dedication; (6) prayer and blessing for prosperity; and (7) conditional blessings and curses to ensure future upkeep.[8] This is a near-exact description of the Exodus account in its entirety. The technology was known to be used by the Egyptians and therefore could have been a skill that the former Israelite slave population would have possessed prior to entering the wilderness. Mimicking known patterns in their own religious practices is highly probable.

In contrast, Julius Wellhausen calls the tabernacle a "copy, not the prototype" of the temple.[9] Wellhausen believed it was a "pious fraud"[10] resting on historical fiction borrowed from the priestly code rendered from construction records of the temple. By Solomon's time no tabernacle, nor holy vessels, nor brazen altar existed. At best, there was a generic tent with the ark or representation of the ark with David. Wellhausen contends that the

---

[7] Kitchen, "Egypt's Impact," 213. "Tutankhamun was found with gold plated solid walled tabernacles around him. . . . Ritual boxes with rings for poles were also found there, resembling the description of the Ark."

[8] Dozeman, *Exodus*, 572; Averbeck, "Tabernacle," 816.

[9] Wellhausen, *Prolegomena to the History of Israel*, 764.

[10] Averbeck, "Tabernacle," 818.

tabernacle reflects a prototype of a "halved" temple, made up during the exile to offer credence to the wilderness wandering. It was made portable to reveal God's omnipresence, even when God's people rebel against him. The priestly authors took what they knew of the temple and translated it into a wilderness shrine under Moses's supervision, with the various terms used in reference to it done so interchangeably due to the various traditions that are wrapped up in them, including pre-temple shrines, the temple itself, and other later synagogue traditions. Another reason for historical non-belief in the taber-nacle is due to the sheer size and weight of the materials along with the lack of people with the ability to pull off the project. It would have consisted of one-and-a-quarter tons of gold, four tons of silver, and three tons of bronze. It is argued that the haste of leaving Egypt would not have allowed the Isra-elites adequate time to gather such quantity of materials.[11] Likewise, during the time of P—the presumed author—these materials would have been traded regularly and could be easily added into the story to make it feasible that the Israelites had the large amounts needed. In 700–500 BCE, the skilled craftsmen, people, and materials necessary would have commonly existed in order to fit into P's imagined 1300 BCE wandering society.

In conclusion, it is not unthinkable that the Israelites set up some sort of worship center in the wilderness. Since scholars generally agree that the ac-count of the tabernacle is attributed to P and is from the exilic or postexilic period, it is, therefore, likely not an early account to the time of Moses. As discussed, the arguments range from it being a pious fraud to a valid historical memory. Yet, "no critical scholar accepts that the account in Exodus is a literal account of the desert shrine."[12] With so many details left out, even if presented as a directive "blueprint," Bezalel would have been able to decide how the intricate details would be incorporated into the final product. Con-sidering the other ancient Near Eastern examples, it can be argued that some form of the biblical tabernacle truly existed, though whether it matched the biblical description in every detail is open to question. There are sufficient parallel accounts and material plausibility for the existence of the tabernacle. The narrative tent was probably formulated out of something more straight-forward and idealized from the time prior to entering the promised land.

In this way, Bezalel and Oholiab would have been allowed some level of poetic license if the general framework was followed. Therefore, how the tabernacle would have actually looked cannot be fully rendered. Neverthe-less, regardless of the physical realities or fallacies as commonly perceived today, the tabernacle's theology is directly tied to the holiness of Israel. In

---

[11] Hillman, *The Torah and Its God*, 226–27. I propose that if there were 400,000 to 1,000,000 Israelites who were allowed to plunder the Egyptians on the way out of Egypt during the Exodus as is commonly thought, it is conceivable that the goods required for the construction of the tabernacle would be plenti-ful. It would require a small amount from 400,000 individuals to collect the material supply required.
[12] Klein, "Back to the Future," 264.

whatever form it existed, its purpose was always a material reminder of God's place among his people and sacred space from which to worship him.

P'S TABERNACLE: BEZALEL THE UNINTENDED PRIEST

Regardless of contemporary arguments as to authorship, the Pentateuch is mostly understood as an edited work and not a piece of literature penned by one person at one time.[13] At minimum, it is considered an assortment of documents written down, collected, and edited prior to compiling the entire volume. The tabernacle narrative, and thus the inclusion of Bezalel as a biblical character, is considered to be a redaction by the Priestly (P) author—either in the exilic or postexilic era, approximately 500–450 BCE—who sought to establish a permanent system of liturgical directives. P is widely considered to have added much—if not all—of the creation and redemption narratives.[14] The political purpose of the tabernacle narrative was to create a place for the priestly class for all time. P separates Moses from the ruling tribe in order to gain or retain power over the people as a God-chosen ruling class. The P redaction demonstrates the importance of the priesthood through leading the reconciliation of all humanity; the inclusion of all Israelites serves as a unifier. Where a wholly Mosaic leadership could create a human idol over YHWH, P's priesthood serves to bring the people into knowledge of and communion with their God from within a created sacred space.

Bezalel possesses the exact, yet greater, qualities of Hiram in the temple. The similarities between a known—or possibly recently known—artisan in Hiram shows P was perhaps superimposing a commonly understood, factual temple account over a re-creation narrative. This offers a literary-historical explanation of the reason for the demise of the Davidic dynasty, the exile, and Babylon. As well, it offers hope that a future glory will come, or has recently come, despite the people's unfaithfulness.

David Baker summarizes P's purpose: "While the Pentateuch was not written as a historical document but as a theological one, its theology is

---

[13] T. Desmond Alexander ("Authorship," 61) summarizes the authorial controversy: "For some, to doubt the long-standing tradition of Mosaic authorship is the greatest heresy. For others, to support unquestioningly the belief that Moses wrote the whole Pentateuch is the hallmark of blind, uncritical thinking." Neither position can be argued with complete certainty. This book adopts the common understanding of a redacted JEDP authorship ranging from approximately 850 to 450 BCE as a collection of various writings by multiple authors. Calling the Pentateuch the "Books of Moses" does not have to mean that he was the author in totality, but instead could be used as a title designating the subject of this section of the Bible.

[14] The Priestly Code (P), built from the El school of authorship, represented the four covenants of Adam, Noah, Abraham, and the Sinai, as the framework. The addition of the law would have fit in with the postexilic era, making JE originate from the divided monarchy, D from the seventh century and Josiah's reforms, and P from the postexilic restoration. In this way, P is a postexilic supplement to J.

nevertheless historically based, and impugning its historicity has theological outcomes."[15] P lived at a time when God's judgment and attack from other nations caused God to abandon the temple (Ezek 10). YHWH's dwelling with man was only temporary. P's primary goal appears to be to re-establish the cultic community of the tabernacle, priesthood, and sacrificial system. Bezalel's—and therefore the people's—obedience in constructing the dwelling place would demonstrate their loyalty to God, a loyalty demonstrated through God's chosen technical artist. P concludes the narrative by presenting the people as blessed after fully obeying the call to build. In this way, P sought to keep the holiness of God as the central focus by presenting the Israelites as a worshiping community.

Benjamin Sommer takes a different view, noting that P always speaks in terms of a tent and without any acknowledgment of the temple ever having been built. P is concerned about what is utopian, not about what is physical.[16] Mark George likewise adds: "The social space of Israel's wilderness tabernacle is one means of creating an orderly, consistent world and cosmos for the Priestly writers, even while they find themselves living in exile, displaced."[17] P sanctions Yahweh's presence in the tabernacle with the special verb *shakan* which is usually translated "to dwell" or "to tabernacle," showing the reality of the tent. God does not live there in the same sense of people living in house. He has a metaphysical "presence" there suggesting that the universe is back into balance. Sommer summarizes the arguments:

> P's silence on the issue of the temple or the sacred city makes it impossible to decide between Kaufmann and Wellhausen on this issue, and other possible readings of the crucial passage exist as well; it is possible that P's tabernacle did not originally stand for any one sacred site but came to represent the Jerusalem temple as priestly tradition developed. But the fact remains that P does not explicitly connect the tabernacle to the Jerusalem temple or even to multiple Israelite temples. It only describes a wandering shrine that is located at the center of the camp, thus suggesting both locomotive and locative understandings of that shrine.[18]

According to P, the tabernacle is also the place from which God's law code is revealed. It serves as the single legitimate place of regular worship for Israelites in the desert. Not only does God approach Israel there, but Israel approaches God as well. The priestly tabernacle is the sacred center, the capstone of the universe where God is constantly and reliably manifest.

To P, God is always present, while still restricting atonement, redemption, meeting, and God's presence to a central spot. In this way, P focuses the

---

[15] Baker, "Source Criticism," 802.
[16] Sommer, "Conflicting Construction," 53.
[17] George, *Israel's Tabernacle*, 72.
[18] Sommer, "Conflicting Constructions," 55.

Pentateuch on the immanence of YHWH through a physical, sacred worship space that is confined yet not limited.

CALLED BY NAME

Selection of the artisans to build the tabernacle was not happenstance; God specifically called Bezalel and Oholiab by name (Exod 35:30).[19] That Bezalel and Oholiab are identified prior to the charge to build is given suggests that they are personally and perfectly equipped for the task, for they were surely not the only artisans present among the people as the call for the others to assist Bezalel and Oholiab shows. In the Hebrew tradition, God's action denotes a single individual "named" to a specific job and commissioned to a specific task. The LXX suggests a favored status of Bezalel with *anakeklemai ex onomatos* meaning, "I have called out for myself by name." In this way, Bezalel, and no one else, was to oversee the tabernacle's construction. Throughout the Bible, God only calls people by name whom he puts into high places as emphasized through the Lord filling Bezalel with the Spirit of God. In contrast to Moses, Bezalel never questions his calling to the project. Bezalel appears to understand the importance of the calling and recognizes he is endowed explicitly with the qualities necessary to complete the assigned task.

Bezalel was likely a skilled craftsman before he received the divine commission, yet God's filling of the Spirit elevates Bezalel's previously held skills to a level required to accomplish this job. The calling of Bezalel was in the past tense: I "have called and filled." He already possessed the skill. These talents were then to be employed in service of the Lord. Whom God chooses to do the work, he also bestows upon the necessary skills for the handiwork both in mind and body, which suggests natural talent and supernatural anointing are not opposites but instead work in relation to each other when used for God's purposes.

Bezalel and Oholiab's contribution is an offering of their whole selves, an internal filling that creates an outward expression of worship. It is an inner ability first. It is a gift from God for a specific purpose as demonstrated throughout the Bible in the same way others are called in service of the Lord. In the surrounding passage, God tells Moses that he knows him by name. Just as Moses was given a specific purpose in service of God's people, so too does God know the "name" of Bezalel for the purpose of assembling his

---

[19] George, *Israel's Tabernacle*, 163. George suggests that because Bezalel and Oholiab are identified before the charge to build is given, this adds to the idea of each one as an individual with a specific purpose. They were not merely the only available artisans with the skills to perform the task. It is a special assignment for select artisans to complete.

dwelling place on earth. This denotes an intimacy between God and the artisans: when an artist acknowledges the truths of God, a higher quality is produced. The artist's calling by name is as an example of God's grace on humanity. God's endowing of their gifts for the sacred work makes them honored servants. In this way, calling Bezalel and Oholiab by name is personal to them in a way not scripturally common.

BEZALEL: IN THE SHADOW OF GOD

Bezalel—"Bezalel, son of Uri, grandson of Hur, of the tribe of Judah"—means "in the shadow and protection of God."[20] Though Bezalel is a lesser known character in modern biblical study, his name denotes that he is special to God, holding a name that suits his overall purpose. His name is an acrostic conjunction of *B'tzelem Elohim*, a metaphysical representation of the "image of God."[21] His father's name, Uri—short for Uriel or Uriah—translates to "God/Yah is my light," making Bezalel, figuratively, the Son of Light. Philo saw this as replication nomenclature in that Bezalel's name denoted that he was a replica of God in the same vein as *imago Dei*.[22] The exact Hur—meaning "free man"—being referred to as his grandfather is unknown, but it could be the same person who judged with Aaron (Exod 24:14), accompanied Aaron on the mountain in the war against the Amalekites (Exod 17:8–16), and held up Moses's arms (Exod 17:10–12). According to rabbinic tradition, Hur held a prominent place in Mosaic leadership as the husband of Moses's sister Miriam and was killed in attempting to stop the creation of the golden calf. This should not be quickly overlooked: P places the exact grandson of the one who defended against worshiping a false idol in place of YHWH as the person who would ultimately create the dwelling place of God among the Israelites. In this way, both Bezalel's namesake and lineage have ties to the sacred worship practices and protection of God's "name" among the people. The

---

[20] John Durham (*Exodus*, 409) translates the meaning of Bezalel's name as "in El's protecting shadow." He suggests this adds a sense of action by God upon Bezalel. In this way, Bezalel's actions in creating the sacred worship space for YHWH are also actions of God himself. Where God is his protection, he likewise presents God to the people for both their protection through corporate worship and the preservation of his memory.

[21] Aranoff (*Torah*, 33–34) suggests there is an alternate method of grouping the letters of Bezalel's name in order to make it mean "God's Onion." This is an interesting suggestion. While suspect, when viewed this way it does support the qualities that an artist would possess. Each layer reveals the various skills. Even if the letters of his name are not to be arranged to create the "God's Onion" meaning, in an oral culture, the combination of letters to sounds could generate a play on words, which in turn could offer a secondary and more in-depth interpretation of his role as well as his personal characteristics.

[22] Durham (*Exodus*, 409–10) proposes that "Ur" translates to "my flame." The translation suggests greater interaction between Bezalel and God, along with suggesting his source of energy and inspiration. Durham refers to Bezalel's role as "the supervising artisan" for the "various media of worship." In this way, Bezalel possesses the fire that also provoked Moses to respond to God's calling to serve his people. Thus, Bezalel becomes an embodiment of the power of God to lead people to himself.

combination of the meanings of his namesake parallels the fact Bezalel is filled with the Spirit of God, *ruah Elohim*. Thus, his name suggests that he is not only in the shadow of God but more so in the direct protection and purposed care of God. He is a literary—if not literal—"son" of God.

Being a Judahite is also an important factor because once the tabernacle is constructed, no other tribe than the Levites are allowed to touch it, carry it, take it down, use it, or enter all areas within it.[23] The service of the tabernacle was explicitly a Levitical job, even if the construction was reserved for those who were spiritually gifted. Descending from the line of Judah places Bezalel in tribal lineage with both David and Jesus.[24] Bezalel most likely would have been viewed as an artistic patriarch, a symbol of worship practice and liturgical structure, to both David and Jesus. Likewise, first century Jews and early Christians all believed Bezalel was a real person. While writing for the Romans, Josephus referred to Bezalel as "renowned." Holding him in high regard presents him as a suitable model of a regular citizen commissioned for the construction of significant masterpieces for the priests and ruling class. Josephus's choice of words would place Bezalel within a class structure important to the Romans, one wherein Bezalel would be worthy of honor but without challenging the ruling class's higher status.

In Christian circles, Bezalel is often called an architect and builder. The original Hebrew understanding of him is as an "artisan and designer." The Greek *arkhitekton* signifies a chief master builder, artisan, or craftsman.[25] Because Bezalel was filled with the Spirit, he was able to create according to both his own will and the power of God, not only due to his possessed skill to perform the task. In the same way, Michelangelo compared himself to Bezalel, seeing himself as a divinely inspired artist. Michelangelo saw artisans like Bezalel—and himself—who create through the Spirit of God as able to overcome Moses's iconoclasm, allowing their creations to be living examples of gospel truth. It cannot be overlooked that Michelangelo was not only the most talented artisan of his day, but understood his role in the service of

---

[23] Though the text does not ever cite Bezalel or his artisans assisting with the tabernacle, it can be argued it is their expertise that would be required to reconstruct and maintain the tabernacle and its belongings throughout the remainder of the wilderness journey. Assuming the tabernacle was carried throughout the desert for the remaining thirty-nine years, normal wear-and-tear would suggest that the skilled artisans would be required in an ongoing basis and not only until completion of the construction project.

[24] According to the biblical text, Bezalel's full lineage from Judah would encompass Judah to Perez to Hezron to Caleb to Hur to Uri to Bezalel. This is most likely redacted to show only the major patriarchs in his ancestry.

[25] Blum, "Vasari on the Jews," 560–61. Bezalel's popularity peaked in the fifteenth and sixteenth centuries. Bezalel's and Oholiab's tabernacle was a common depiction. "For Vasari, Bezalel, the first Jewish artist, stands at the beginning of art history in the same way that Moses stands, in the Hebrew Bible, at the beginning of all written Revelation." Blum notes that Giorgio Vasari viewed the Jews as "protagonists" or fans rather than active in the development of the art, while Bezalel was the first to define religious art because God left artistic construction to him. Vasari states, "the art of design . . . was bestowed by the mouth of God on Bezalel . . . for no other reason than to bring the people to contemplate and to adore them."

God's dwelling place in a parallel to a biblical character. In this way, modern church technical artists are the Michelangelos of modern church media. Their creations, when made as a calling to their namesakes as artisans, are hence "in the protection of God," able to share "God's light," and "free man[kind]" from the cultural barriers that promote false worship.

OHOLIAB: THE ASSISTANT

To fully understand Bezalel, one must understand his assistant Oholiab ("Oholiab, son of Ahisamach, of the tribe of Dan"). Oholiab means "The Father-God is my tent," and uses the same word that is found in "tent of meeting" from Exodus 27:21, which suggests a care and protection of God through a physical meeting place. His name implies a divine covering of protection. Oholiab's father's name, Ahisamach, means "my brother offers support" or "my brother has leaned/sustained [his hand for support]." This is a fitting surname for the man called to perform the tasks as Bezalel's assistant. Oholiab carries a special sense to it: it is made up of *'hl-'a*, which denotes both tent and father. The use signifies a divine Father rather than human father. Therefore, a dual meaning is present; not only is God a physical protection as the "tent" but he is also a true person in the form of the divine "Father." Protection is found in the tent that is the Father.

Oholiab is from the northern tribe of Dan, which would have been viewed as an honor. That Bezalel's assistant is from tribe of Dan is significant because Dan—the person—is the first of Jacob's sons, and Dan—the lineage—produces the famous judge Samson. The tribe of Dan was known for its leadership roles and assisted heavily in the administration and construction of the temple. From a literary perspective, the inclusion by P of an artisan from the tribe of Dan suggests that there was leadership, strength, and wisdom in performing the work being untaken. Thus, the conclusion can be drawn that the skills required to perform great artistic feats in the creation of worship spaces is not attached to any one tribe but rather that all tribes and peoples are endowed with capable artists.

Like Bezalel, his name is fitting to his purpose. Oholiab is the tent-making apprentice who comes alongside God's "shadow." His name is an obvious play on words because he is literally building a tent for the Father.[26] The connection between his name being the actual building of a tent and God being found in a tent should not be overlooked: the Hebrew understanding pictures a name that is not of a physical dwelling alone, but metaphysical. Whereas Bezalel was the divine architect and artisan, Oholiab knows where refuge is

---

[26] According to Sarna (*JPS Torah Commentary*, 200), names including a derivative of *ohel* (tent) were common in Phoenicia and Arabia.

found. His name is used in the Old Testament's chapter on deliverance of God's people, and the first attempt at God's dwelling among them through building the physical sacred space of God and the shelter for his people. Like with Bezalel this should not be glanced over lightly. Artists who construct have a special purpose when creating worship in God's modern tent, the church. They create the place of refuge for God's people at the dwelling place of God.

TABERNACLE ARTISAN WORKERS

Bezalel and Oholiab are the central leadership team in tabernacle narrative. Likewise, "skilled artisans" with "stirred hearts" from within the people of Israel play a central role (Exod 35:35–36:1). Kidwell explains:

> This literary celebration of willingness stands as a strong repudiation of the forced labour that is described at the beginning of the Exodus account. Those who are summoned to undertake the work (Exodus 36:2, cf. also 25:21–2, 26) are volunteers: '*kōl 'ašer nesā 'ô libbô*' ('all those whose hearts were stirred'). The LXX is more straightforward, with '*kai pantas tous hekousiōs boulomenous*' ('all those who willingly desired'). In the Hebrew, the verb *ns'w* (lit: 'to lift up') is used figuratively here, thus it may be rendered as Walter C. Kaiser suggests: 'the heart 'lifts one up' thus inciting action.' As the LXX translator emphasizes, with the use of the verb *hekousiōs*, human participation in this enterprise is necessarily free inasmuch as it is communal.[27]

The Israelites go from building storehouses and cities for Pharaoh to erecting a house and dwelling place for God. The text suggests God's bestowing the Spirit, skill, wisdom, intelligence, and understanding are traits shared among various people and not ascribed solely to two particular individuals from the tribes of Judah (Bezalel) and Dan (Oholiab). The text states that all who were skilled were called to perform the tasks for building the sanctuary. Both men and women were involved.[28] In Israelite society, the women would have prepared the fabrics and wool. Therefore, in Exodus the description of the priestly garments would most likely have been performed by skilled women, with archaeological evidence for this dating back as far back as the Neolithic period. This is significant that God appears to show no discrimination toward those who are both called and able to serve the critical task of constructing his dwelling place.

---

[27] Kidwell, *Theology of Craft*, 40.
[28] Women are mentioned in the construction narrative, however, not in the instruction narrative. This is significant because the earlier commissioning narrative is Moses's instruction to Bezalel. The later construction narrative is Bezalel's reliance upon God's instructions to him. In the latter, Bezalel finds favor in the assistance of women in producing the worship space.

The passage indicates it is not the priestly class who receive the Spirit to complete the tasks but "all" the "skilled" workers, leading to the glory of the Lord approving of their work, filling the space, and transforming it into God's sacred dwelling place. The designation of "all of them master craftsmen and designers" signifies that amateurs were not involved, but only professionals gifted and guided by God, and that 36:1 demands the builders also be gifted by God and perform their craft according to divine plan. The people could not choose to be part of the work unless fully qualified; they had to be experts. Here, professionalism is a statement of quality and not necessarily employment status. Completing a feat of divine instruction requiring a specific skillset. Likewise, the artisans could not modify God's plans; construction must be according to God's requirements. Including others in construction narratives is not abnormal in ANE literature; however, it is unique that they had an internal response to participate. They were "stirred" and "inspired," and the artisans most likely had the necessary skills sharpened in order to complete the tasks.

A social aspect is also involved, one which makes it a corporate work incorporating the entirety of the Israelites assisting their commissioned leaders. It took a coming together of the entire society—whether through donations of goods or assembling them—to bring the project to completion. Israelite non-artisans were also involved in the project through the giving of freewill offerings, giving so abundantly they exceeded the need (Exod 36:5). The people responded, giving commodities of gold, silver, bronze, and linen from their plundering of the Egyptians: the people gave until Bezalel informed Moses the supplies were sufficient (Exod 36:4–7). Bezalel is the chief administrator of the worship production budget, understanding what was sufficient for creating the sacred space according to God's precise plans. Because the contribution of materials came from the congregation, everyone had the pleasure of contributing in some tangible way to the art of the sacred space by being moved by God in some form, either artistic skill or tithes. Giving was not a chore but rather a willing addition to the solidarity of the community and the call for unity in purpose.

TABERNACLE WORSHIP AND ARTISTRY

The tabernacle is referred to by eight different variations of two main words: *mishkan* and *'ohel*. In practical terms, all variations are synonymous and are applied to places that serve a higher purpose—a sanctuary—not merely a tent: (1) *mishkan* (dwelling); (2) *mishkan* YHWH (the dwelling of the Lord); (3) *mishkan ha-'edut* (the dwelling place of the testimony); (4) *'ohel mo'ed* (tent of meeting); (5) *mishkan 'ohel mo'ed* (dwelling place of the tent of meeting); (6)

miqdash (sanctuary); (7) *ha-qodesh* (holy place); and (8) *qodesh ha-Qodesh* (most holy place or holy of holies).[29] It is a visible representation of God's spoken promises. The tabernacle is not just about Bezalel and Oholiab: the whole of Israel was invited to enter into worship through the tabernacle experience. It became a symbol of the people's heart for God, and God's heart for his people.[30] At the center of worship was the ark of the covenant, encompassing all of God's promises to his people, with the pillar of the divine Shekinah symbolizing the precise placement of the ark and mercy seat, and thus the exact location of God within their midst. The mobile sanctuary had three levels of holiness, from the court to the outer room to the holy of holies. As the graduation becomes more sacred, the value of the materials and level of skilled workmanship is likewise increased: from bronze to gold, from plain weaves to woven patterns.

Because the details provided are not sufficient to be used as architectural plans, it can be argued that the author intended the text for a wider audience, not just those with the technical ability to understand the minutiae. To a layman the text would be viewed as moralistic rather than as specific to a particular group. Religious-rite ancient Near Eastern artisans like Bezalel and Oholiab were viewed as the mediators between god(s) and man because they created the objects to be worshiped, which in practical terms was the god itself. Christ John Otto writes:

> The artisan had a unique role. To a technologically primitive society, a person who could create things seemingly out of nothing was believed to have semi-divine qualities. A person who could take metal and shape it over a forge, or take wood and carve it into the shape of an animal or person (or god) was seen as a person who was making invisible things suddenly visible. . . . The artisan[s who] could seemingly bring things out of heaven . . . began to be seen as mediators between the gods and men.[31]

Three kinds of artistic style represented in the tabernacle: abstract (nonrepresentational), representational, and symbolic. Symbolic art is "the use of physical images to stand for a corresponding spiritual reality."[32] Examples include the golden altar for atonement, the golden table for God's provision, water for cleansing, lampstands for illumination of God's revelation of himself. Christians would suggest that such symbols today might include the bread and wine at the Eucharist and water of baptism. Representational art "portrays the objects of the physical world in a recognizable form . . . [which]

---

[29] Rothkoff, "Tabernacle," 418.

[30] Leithart, "Making and Mis-Making," 313. Leithart proposes that the Israelites were released from Egypt in order to go into the wilderness to hold a feast to Yahweh. That implies their mission is not complete until the feast is held, which offers explanation to the overwhelming details in preparation of the feast. The preparation included the creation of God's sacred space among his people.

[31] Otto, *Bezalel*, 81–82.

[32] Ryken, *The Liberated Imagination*, 54.

imitates the forms of life as they pass before our gaze."[33] These include adding inscriptions of physical items like flowers, pomegranates, almonds, lions, oxen, and palm trees. It shows that all of God's creation is to be admired and is suitable for artistic embellishments. Abstract art is "art that represents nothing beyond itself. It is simply a pattern or design that is complete in itself."[34] The combination of real to figurative, plain to ornate, work together to give the feel of reverence in themselves. Importantly, God gave many details, but not enough for us to confidently know what the tabernacle physically looked like. For example, the tabernacle contained representations of items that that never existed as currenting like the blue pomegranates, which do not exist in reality. This suggests, therefore, that some amount of artistic license must have been allowed. The description of what to do within the tabernacle is in the text, but the exact style is not specified.[35] In fact, the modern image is skewed by early scholarly attempts to recreate a "possible" option, like those of James Strong (1888, 1893). All three of these artistic styles are and can be incorporated by technical artists through visual, auditory, and scenic embodiments.

Otto concludes: "God lavished on Bezalel an abundance of creativity. And although the Tabernacle would be dedicated to God, the Tabernacle was for the benefit of the people."[36] Thus, God allowed the worship of him to exhibit a level of creativity endowed to Bezalel and his artisans that extended beyond the spoken instructions offered to Moses. In modern art, this would be equivalent to artists taking poetic license with their works. In this way, contemporary church technical artists ought to take creative license in their craft insofar as the principal structures and purposes (i.e., doctrinal stance, or the church mission statement) are not infringed upon.

MOBILE REMINDER OF WORSHIP

Tabernacle portability made it the most ingenious plan and workmanship of its day. It allowed a wandering people a closeness and connection to their God not previously seen in history. It served as a mobile reminder of the need to worship, keeping God at the center of both their camp and their travels. That the command to build the tabernacle comes after the golden calf

---

[33] Ibid., 55.

[34] Ibid., 56.

[35] I propose a modern example: Tell a classroom of thirty elementary schoolkids to draw a picture of a red truck exactly twenty centimeters long. The final products would include thirty very different representations of a red truck, yet all of which would contain exactly the same predefined characteristics. The tabernacle can be viewed in the same way. The plans given are interpreted according to the loci of the artist called to recreate the plans.

[36] Otto, *Bezalel*, 61.

rebellion shows concession to Israel's need for a central place of worship of YHWH and that it was not an agenda from humans, but from God himself. The people's disobedience created the need for a permanent space to demonstrate true obedience. The tabernacle served as the created space of sacred meaning in which the people of Israel are able to dwell with God where ever they may be on their journey. "God is both with them and for them."[37] It offers a tangible representation of God, his perfection, his dwelling, and space to receive worship. As a moveable Sinai, being dedicated on the first anniversary after the Exodus from Egypt, it serves as a restoration for the people and a new world for them to worship God. Where they were previously a people on the run, they are now a people dwelling with God among them, even while displaced from their promised land. Being portable means that the people of Israel had to continuously "rebuild" their recognition of Yahweh and his presence within their midst at every step along the journey. For modern church technical artists, there is a direct parallel to the current "satellite church" phenomenon (known as the "roll-in-roll-out" or "set-up-tear-down" church). A large number of churches purposefully—though most likely for monetary reasons—do not meet in a permanent location. Instead, they rent school theatres, gymnasiums, community centers, or even local parks in order to temporarily "hold church." They turn a building into a sanctuary for a controlled period of time. Each week the congregation establishes their own mobile reminder of Christ in their midst within the same locations where their daily lives revolve throughout the rest of the week. In this way, the technical arts teams utilize the same equipment that may be used for a musical theatre or drama class Monday through Friday for the purpose of bringing God into the congregation's midst, dwelling and sharing his identity, if only for a few hours on Sunday morning.

The tabernacle had to remain holy. The Levitical call to be encamped around the tabernacle offered protection of its holiness. God commanded Bezalel to use only the best materials and workmanship in its construction, just as God demands only the best to be brought as an offering (Exod 12:5, 25:2; Lev 22:21; Prov 3:9). When YHWH fills the tabernacle with his glory, he confirms that it was prepared with the quality and exactness he requires. With that, future generations are reminded that where they worship is exactly as God expressed (Exod 39:32) and finds acceptable.

Making a tent was not unique to Israel, but the combination of items with a specific purpose and specific builders in Bezalel and Oholiab—which in total produce Israel's social space—mark it as unique to the Israelite's identity. It is a symbol of their steadfastness. The Bezalel narrative—surrounding the story of impatience and rebellion—offers the Israelites an unsurpassed stability in the midst of transition and insecurity. The tabernacle construction

---

37 Fretheim, "Because the Whole Earth is Mine," *Interpretation.*

account itself is interrupted by the Sabbath law, which emphasizes the importance for obedience and using God's created earthly dwelling place as a place of rest and holy time, thus serving as a reminder for undistracted worship.

As slaves, the Israelites would have had no wealth. Because the items given for the tabernacle construction are those received from the Egyptians—and the commissioning and call to give would have come only months after leaving Egypt—the people would have no problem offering them for the construction of a sacred place to worship their deliverer. Seeing the offering as only months removed from the exodus and not forty years—as many commonly envision the narrative—places the commands into realistic perspective for those involved. Likewise, with estimations of the exodus including between 400,000 and 1,000,000 Israelites, the total number of precious items for the tabernacle would have been a small percentage of the overall pillage from Egypt. The passage suggests that even though the offering was voluntary, and that the goods were received because God allowed the Israelites to plunder the Egyptians, YHWH still viewed the offering as an act of worship. God provided for himself the precious items—and not just that which was readily available—yet the people must offer the materials freely. Thus, every action of the tabernacle, from the collection of goods, to construction, to worship within it, demonstrates that it is a mobile reminder for both worship and as a continual voluntary offering of themselves due to the presence of God among them.

CREATION MADE PHYSICAL: GENESIS REWRITTEN

The tabernacle narrative holds distinct similarities to the creation narrative. Since P is thought to be the author of both, it can be said that Bezalel is the earthly example of the heavenly Father. P presents Bezalel and Oholiab as possessing the same power in creating as God possessed over creation. P utilizes the tabernacle as the natural bridge between the delivered-but-not-yet-arrived people awaiting their promised land. Thus, the tabernacle narrative can be viewed as a microcosm of cosmos, as demonstrated through the literary parallels to both the creation narrative in Genesis and the presumed "heavenly pattern."

Exodus ends where Genesis begins. For the first time since the fall, God is able to dwell with humanity. God's plan was always to dwell with his people as the only one worthy of their worship. Unable to keep their faithfulness in the garden—humans living in God's chosen home—God brings the garden to them, creating a place for himself among the people through the tabernacle. It is "Heaven on earth."[38] In the tabernacle construction the builder is not God, but an earthly replacement who parallels God's qualities as presented in creation. J. Richard Middleton explains:

> As overseer of tabernacle construction, Bezalel is filled with "wisdom," "understanding," and "knowledge," precisely the same triad by which God is said to have created the world in Proverbs 3:19–20. To this is added that Bezalel is filled with "all crafts" or "all works," the very phrase used in Genesis 2:2–3 for "all the works" that God completed in creation. Therefore, not only does the tabernacle replicate in microcosm the macrocosmic sanctuary of the entire created order, but these verbal resonances [also] suggest that Bezalel's discerning artistry in tabernacle-building images God's own construction of the cosmos.[39]

The construction narrative holds many parallels to the creation account: God saw; Moses saw. God had made; Bezalel had made. The same precious stones found in creation are the first and last listed in the tabernacle account. God: "Behold!" Moses: "Behold." God finished all the work, while Bezalel finished all the work that God commanded. The Spirit of God placed into Bezalel is the same Spirit of God that hovered over the waters. Humanity was created for making, so the artisans made. God blessed creation; Moses blessed the creators in a way that parallels God's declaration that the creation was "very good." In this way, declares Philip Ryken: "The calling of these artists reflects a deep truth about the character of God, namely, that he himself is the supreme Artist,"[40] and they serve as his earthly representations. The tabernacle is an extension of God's creative works. Therefore, the tabernacle is a complete and ordered environment paralleling the perfect pre-fall world of creation. It is offered for the community of faith to commune with God, just as God was present in Eden. It offers the Israelites a continued presence of God in all his glory.

Talk of being "filled with the Spirit of God" may reflect language used in Genesis 1:2.[41] Where there is creation, the Spirit resides. Tom Nelson goes so far as to suggest that this is an OT reference to the Third Person of the Trinity,[42] which modern Christians could easily attest to, even though most likely not possible to be understood as such by P. Bezalel mimics the creative role of God, using natural materials in a god-like way, mediating the place of worship for the person of worship. Therefore, rather than viewing the tabernacle as a creation narrative, it is better understood as a "re-creation"

---

[38] Fesko, *Christ and the Desert Tabernacle*, 116.
[39] Middleton, *The Liberating Image*, 87.
[40] Ryken, *Art for God's Sake*, 103.
[41] Hebrew word *ruach* (and Greek *pneuma*) can mean "spirit," "breath," or "wind." In Hebrew it also means to breathe through the nose.
[42] Nelson, *Gifted for Work*, 146.

narrative.[43] Richard Hess furthermore suggests that having two builders present in the construction parallels the "we" found in Genesis,[44] with P in both narratives suggesting there are two man-like forms of God. The "image of God" is represented in the plural in Genesis, just as there are two builders.[45] Outside of the "we" in Genesis, God is not given other distinct characteristics according to the modes and items of creations like Bezalel and Oholiab are. Though making a direct correlation between God and Bezalel and Oholiab must be made with caution, the parallels are noteworthy nonetheless. This suggests that when God's people perform their creative works for communing with God, the Lord is present in the work like he was present at creation.

A form-critical examination of the text finds similar parallels: The passage is YHWH's sixth of seven speeches—each beginning "The Lord said to Moses"—with the ultimate purpose both in Genesis and Exodus centered around God's perfect dwelling with humanity[46] In Genesis man is created on the sixth day of seven days. In the sixth of seven speeches in Exodus, man once again dwells with God.[47] Bezalel and Oholiab—the focus of the sixth speech—demonstrate humanity's dominion over creation: God commanded man to tame the earth, using his knowledge and wisdom to rule over it.[48] Likewise, God commanded humanity to create God's earthly universe within the confines of the tabernacle. Thus, God is the creator of creation and tabernacle: They are both divine activities.[49]

The tabernacle is the "terrestrial objectification of a celestial image," suggests Nahum Sarna.[50] At Sinai YHWH directed Moses to prepare a building

---

[43] Leithart, "Making and Mis-Making," 314. When viewed in light of the NT, just as Christ came to point mankind to himself and the Father, restoring the broken relationship between God and humanity, Bezalel re-creates God's place among his people, directing them to God in worship.

[44] Hess, "Bezalel and Oholiab," 171. Making this parallel connection is a stretch, but notable nonetheless insofar as P is understood as the common author between this text and the creation narrative. In this way, common themes could be carried over from one narrative to the other. The depictions have similarities with the Spirit as a force working through God's "we" identity in Genesis and the Spirit being the guiding force for both Bezalel and Oholiab.

[45] I believe this may be a dangerous leap because Richard Hess appears to be taking two similarities and forcing correlation where there is no direct proof of it other than connected authorship.

[46] Leithart ("Making and Mis-Making," 314) builds on the seven-day narrative but concedes that any attempt to draw exact parallels to each day of creation will fall short. The similarities due to common authorship are notable and demonstrate the chiastic nature of the Genesis-Exodus narrative.

[47] Likewise, on both the seventh day and the seventh speech, God rests. As well, the construction narrative concludes with Sabbath recognition: the people rest after completion.

[48] Kearney ("Creation and Liturgy," 384) suggests that the two tabernacle narratives separated by the golden calf account that extends from Exodus 25 to 40 follows a popular Egyptian and Babylonian creation motif of creation (Exod 25–31), fall (Exod 32–33), and restoration (Exod 34–40).

[49] George (*Israel's Tabernacle*, 183–87) acknowledges one significant difference between the two narratives. In the creation narrative God performed the action through speech, while the tabernacle is built through the intermediary person of Bezalel who has to be filled with God's Spirit. It could be argued then that the tabernacle is not a work of God, but of man alone. See it as such would be problematic because, in both, God has his hand in the method used, whether it be through his speech or his Spirit empowering the artisan. Likewise, God speaks the instructions to Moses, making it his commissioned work. In the same way, in the NT, Christ performs action on behalf of the Father as intermediary.

[50] Sarna, *Exploring Exodus*, 200.

for his worship according to the "pattern" (Exod 25:9; 25:30; 26:30; Heb 8:5) as an earthly representation of the heavenly. It is the pattern God desired for his people to follow in the garden and given in redemption to a fallen people. Its crafted order symbolized God's spiritual significance to his people in the worship experience and creation of it. Just as God was particular in creation, so too he commands exactness in his creation of the tabernacle, making it the counterpart to the heavenly shrine. Bezalel does not create what is his own imagined object, but that which is endowed by God and placed into him through God.

A balancing act is taking place here. Even in human creations, God remains active and in control. When creative works are performed for God's dwelling, could it then be suggested that the Spirit of God is necessary in order for acts to be performed according to God's "pattern"? That is a difficult leap to take because non-believers can likewise create great works usable for the worship and dwelling of God. However, the text argues that God is at work, possibly even when the artisans are unaware. In this way, what technical artists do on earth—when performed in agreement with the character of God—can be said to parallel the heavenly.

PRINCIPAL THEMES

God not only gave the instructions for building the tabernacle, but also perfected the ability to do so through specific qualities indwelled within his chosen technical artists. The text illuminates six themes concerning the characteristics held by the artisans: (1) the Spirit of God; (2) ability, skill, and wisdom; (3) intelligence and understanding; (4) knowledge and artistic expertise; (5) craftsmanship and know how; and (6) teaching capacity. According to the text, a technical artist must possess all six qualities, with ability, intelligence, knowledge, craftsmanship, and teaching, all falling under common grace. The filling of the Spirit makes the work distinctive.

THE SPIRIT OF GOD

The Spirit of God empowers Bezalel and company to work up to God's standard, perfecting their pre-existing artistic attributes. For Bezalel, it is not merely a gift of common grace but "special grace."[51] Ability, intelligence, knowledge, and craftsmanship were not simply "zapped" into them but rather were developed over time and perfected by the Spirit. The Spirit

---

[51] Veith, *Gift of Art*, 22.

supernaturally gifted them for this particular calling, making the work stand out as something special to behold, more than just a creation made by talented artists. "The example of Bezalel suggests that the biblical prophets . . . always possessed the technical skills we see in them; God/the writer is simply making us aware that the true source of those skills is God's Spirit, not just physical abilities."[52] Likewise, the word used for "filled" is in the *pi'el* imperfect tense, which denotes something that happened in the past but also continues into the present and future. Therefore, the completion of the tabernacle did not halt their ability to create after its commissioning. Instead, their abilities could continue to serve God's purposes, for example, in the continual setting up, tearing down, and repairing of the tabernacle as the Israelites traveled throughout the wilderness.

As the primary gift given to him, the Spirit becomes the most important of all six endowments. Jay Wright summarizes:

> When the Israelites spoke of the Spirit of Yahweh, it was often simply a way of saying that God himself was exercising his power on the earth, either directly or, more commonly, through human agents. The Spirit of God is God's power at work—either in direct action or in empowering people to do what God wants to be done.[53]

When some people in the Old Testament were said to have the Spirit of God, it simply meant that they had a God-given ability or competence or strength to do certain things for God or for his people. God's Spirit empowered and enabled them to do what had to be done. Bezalel and Oholiab are special in the sense that they had the indwelling of the Spirit, and not simply a temporary empowerment; this is an important distinction, and one that is unique to God's technical artists in the OT.

The "Law of First Mentions" signifies that when a biblical characteristic is first introduced, the qualities associated with it establish how further mentions should be viewed. Assuming that the Spirit is first used by God in creation and then first used by humanity in the creative, artistic, technical construction of the tabernacle, then it can be argued that technical artisans possess a special characteristic possible for establishing commune between God and humanity when employing their creative talents. The Spirit later in Scripture—namely the NT—would carry the same power of creation, knowledge, ability, and wisdom necessary for understanding the nature and worship of God. That this is the first time the filling of the Spirit is used in the Bible through humans demonstrates the importance of the tabernacle construction to God as well as the importance of personally selected artisans.

Many people today—particularly in the discussion of artistic ability— equate "being filled" with some otherworldly or bohemian experience. In the

---

[52] Willis, "May You Be Like Bezalel!" 111.
[53] Wright, *Three Essentials of Biblical Worship*, 36.

Bible, however, it is simply being gifted from God the ability to achieve his desired end. Being filled with the Spirit always has an outcome that serves God's purposes. The skills manifest inside believers only reach their potential through the Spirit. The otherworldly influence is that of God rather than self-examination or self-expression dictated by emotional outpouring. It is a dangerous line to walk to speak of an artist as being "inspired." It is common contemporary artistic speak, yet Bezalel's inspiration was a different inspiration that should not be confused with some alternative spirituality that connects them to an otherworldly feeling. For Bezalel and Oholiab, it is strictly in response to God's calling and direction. The Spirit Bezalel possessed demonstrates he was a person of faith. It was not a certain "feeling" he possessed, but it was God as Spirit dwelling inside him, guiding his work. In this same way, the Spirit is not used for the usual priestly realms of prayer or prophecy, but for artistic means through physical materials. Bezalel and his craftsmen were functionaries who in turn fulfill a special priestly purpose in the completion of the assigned task.

In other words, the work of the tabernacle included two actions: (1) God giving the instruction; and (2) God empowering humans with his Spirit to complete said action. The Spirit in the OT can take people who were already gifted and cultivate their talents to their utmost potential. Jeremy Kidwell calls it a "craft-pedagogy,"[54] and John Levison an "endowment."[55] Through Bezalel, the Spirit transcends into the physical world to enable human skills for accomplishing tangible results. In a majority of biblical texts, the Spirit speaks with a prophetic voice, but this text demonstrates that it works in physical talents and skills as well.[56] All gifts come from God; but there is a special calling for those filled with the Spirit to use God's gifts for his earthly, physical, and artistic purposes. God wants Bezalel and his assistants to do something with the filling of the Spirit: do something for him and for the people of God.

SKILL: ABILITY AND WISDOM

Skill refers to an innate talent, a practical ability, the hands-on aptitude to perform a task, and the knowledge gained through experience. The Hebrew

---

[54] Kidwell, *Theology of Craft*, 36.

[55] Levison, *Filled with the Spirit*, 38, 62–63.

[56] Enns ("The Spirit's Ministry in the Old Testament," 261) adds that in the OT the Spirit does not seem to have anything to do with the spiritual condition of the person (e.g., Saul). While, in the NT one must be a believer in order to receive the Spirit. The endowing of the Spirit is the sovereign work of God for a specific task. In the OT God often gives the Spirit to someone regardless of the initial call to follow God. The Spirit in the OT is temporary, yet in the NT it is permanent. However, there is no reason to believe that the Spirit is temporary in the case of Bezalel and Oholiab.

word *hokma*, translated as skill or ability in English translations, can also be interpreted as "wisdom." In this way, *hokma* moves beyond simply possessing ability to knowing how to rightly utilize that ability to fulfill its intended purpose with excellence. The Bible presents wisdom as the greatest of all qualities (1 Kgs 3:9; 2 Chron 1:10; Prov 1:7; 4:6–7; Jas 1:5) and is possibly why the author placed it as the first of the human characteristics. It is a "working wisdom."[57] *Hokma* is better literally translated as "wise of heart," which is important, because in the Jewish culture, the heart was considered the "seat of intelligence" and possessed a practical sense of conduct and moral behavior.[58] Skill thus carries a moral value. It is not enough to exercise one's abilities, but rather the artisan must use his or her skills in a way that conforms to the ways of God. Bezalel is not given wisdom for spiritual activities but for the task of building; he is not given the Spirit in order to evangelize or preach as normally found in Scripture. This opens the possibility that wisdom and ability endowed by God are available for ordinary tasks, specifically those focused on serving God.

Hokma is the gift to understand what is needed to complete the instructions, the discerning talent to solve the complex issue of completing the task, and the experienced hand needed to perform the labor. This explains why Moses could not perform the task. Likewise, being called or given instructions is not sufficient; a technical artist must also possess the physical ability and discernment to utilize the ability properly. It is a "divine skill,"[59] noting that the Spirit made Bezalel's ability wise. It is a skill that encompasses a wide range of artistic disciplines. Aranoff cites Rabbi Scherman's observation:

> [That Bezalel] mastered the wide array of crafts needed to build the Tabernacle was remarkable, if not miraculous, for the back-breaking labor to which Israel had been subjected in Egypt was hardly conducive to the development of such skills. . . . God showed Israel that He had not merely redeemed them from slavery, He had endowed them with the capacity to serve Him beyond their ordinary human potential.[60]

God thus used the Egyptian enslavement to develop the tangible skills necessary and the Spirit to grant Bezalel the wisdom to accomplish the task. History thinks of Moses as building the tabernacle, and even though he takes the credit for it in Exod 40:16, Moses builds nothing. All tasks are executed by Bezalel, Oholiab, and their assistants. They alone possess the necessary skill, wisdom, and heart. Nowhere in the Pentateuch is Moses described as wise,

---

[57] Kidwell, *Theology of Craft*, 63.

[58] Ryken, *Art for God's Sake*, 89–93; Ferretter, "The Power and the Glory," 131; Sarna, *JPS Torah Commentary*, 178, 200. Sarna suggests the phase "granted skill to all who are skillful" comes from Dan 2:21: "He changes times and seasons; he removes kings and sets up kings; he gives wisdom to the wise and knowledge to those who have understanding."

[59] Hess, "Bezalel and Oholiab," 162.

[60] Aranoff, *Torah*, 32.

even though it is a distinct quality stated of Bezalel and Oholiab.[61] What was learned in Egypt could be used for the people of Israel and the dwelling place of their God. In this way, even though the technical skills necessary for serving God may have been developed before the calling to use them for his purposes, the text suggests that God was always involved with the development of those skills prior to the artisan being aware of the need for them.

INTELLIGENCE: ABILITY TO PROBLEM SOLVE

Bezalel receives intelligence and understanding, *tebuna* in the Hebrew—the theoretical, practical, problem-solving capacities to bring the commissioned art project to life. Bezalel had to become an expert in each individual aspect of the whole project as YHWH presented it. Intelligence denotes that when an artisan works, it is not only with his hands but his mind: he or she must be able to accomplish the task with reason and rationale. The text suggests that possessing intelligence indicates there is a special quality necessary to perform the craft well. It is not something irrational but rather it requires knowledge of the world, materials, life, and science. Intelligence, for example, is the ability to know how the materials will react with one another and which ones will work best for a given situation. The instructions are useless without the ability to translate those plans into the physical working of the materials and knowledge of how they come together to create the final product. Bezalel had to understand the science behind the art; it is a "divine science."[62] Examples include knowing how metal cures, how acacia wood weathers, and how the connection points would handle continual setup and teardown. For contemporary technical artists, this equates selecting an audio console for the church. While there are many on the market that could do the job, intelligence informs the technical artist as to which tool is best suited for his or her individual situation, and why. It considers all critical checkpoints like channel count, I/O, processing, user experience, and budget. Intelligence advises of the best decision to accomplish the task.

Having the ability to create is not enough. There must also be understanding, reason, and common sense. Artists are to use both the right and left brain, engaging the whole mind. Where an artistic creation engages the imagination, it must also engage the intellect. The Spirit gave them practical, clear, and rational capacity to build. All great artists possess a level of genius: they are able to evaluate the artistic works of others, appreciating others' work and be able to add a qualitative value to it. Intelligence allows artisans to problem solve, taking the abstract and making it understandable.

---

[61] Hamilton, *Exodus*, 601.
[62] Aranoff, *Torah*, 48.

KNOWLEDGE TO UNDERSTAND PURPOSE

Beyond working with their hands, Bezalel and his assistants had to work with their minds. The had to recognize the purpose behind their creation. Their task was not an assembly line of pieces where the beginning of the chain is isolated from the end. Rather, the text suggests that Bezalel and Oholiab had to distinguish the greater purpose of the project rationally, while fully comprehending the ideals their art would portray, the history and tradition of it, and how its meaning would shape the culture politically, spiritually, and sociologically. They had to understand how their creation would influence the Israelite worldview.[63] Knowledge (*daath*) gave Bezalel the ability to discern how each layer affected the whole and how the whole would inform the Israelite worship practice.

Bezalel had the ability to work in various mediums. Where artists often become specialists in one narrow field, Bezalel's knowledge and skillset were wide ranging, encompassing multiple constructive crafts. It could not be limited to one expertise or one artistic medium. He had to discern how each medium would be perceived and the meanings portrayed by his choices. Because he was given poetic license to design, his choices had consequential meaning. Bezalel had to understand that it is not only a creative ability that makes their works great. He must become an expert in his craft. While some were obvious, like the gradations from bronze to silver to gold from the outer courtyard to the holy of holies, others held emotional or implicit impetus, like the color schemes and woven patterns. Philo referred to Bezalel as the "symbol of pure knowledge"[64] because of the wide-ranging knowledge he had to possess to accomplish such a feat. Knowledge allows one to identify the steps, methods, and outcomes of the craft.

There is a duality of physical gifting with regards to the purpose of the gifting: (1) one in being a human artist; and (2) one in being filled with knowledge of God's desire for the use of it. In other words, all humans can be artists, but only those with "divine" knowledge are able to use it for its intended telos: the worship of God. For modern technical artists this could equate to how gain structure and microphone proximity affect impact the audio channel levels and intelligibility, or how varying intensity levels of red, green, and blue light generate scenic effect.

---

[63] Veith (*State of the Arts*, 111) indicates that in modern society many artists believe they gain knowledge through a need to experience life. However, from a Christian worldview this belief is merely an excuse to sin. He suggests that knowledge of one's craft is gained through experience gained over time due to practical participation in the artistic mediums, learning how they react alongside other mediums.

[64] Grintz, "Bezalel," 557.

Baruch Levine indicates that there is a distinction between the two tabernacle narratives in that the detailed directions are flipped: "In [Exodus] 25–27:19 the order is ark-tabernacle, and in 35–39: tabernacle-ark."[65] Likewise, the order given to Bezalel in Exodus 31 is the opposite order that Bezalel performed the constructive tasks. Even though this is possibly for literary or chiastic style, it could be more meaningful. It demonstrates that when Bezalel is commissioned, he alone would know the proper order of how to complete the project. Whereas Moses would explain the scope, Bezalel would work out the details. Moses shared God's command; Bezalel possessed the knowledge to translate that into YHWH's tent of dwelling. Whereas a slave would possibly follow the directions exactly as given, Bezalel, the artisan, took the instructions and discerned how to properly integrate them.

CRAFTSMANSHIP

Adding to the definition of craftsmanship from chapter 1, craftsmanship (*melakah*) here refers to an artist's technique. It is the difference between simply creating and creating well. It is qualitative in nature. Craftsmanship is the difference between executing poorly and fashioning a masterpiece. Craftsmanship allows the medium to take aesthetic form, demonstrating humanity's dominion over the physical materials—over creation itself. Christ John Otto suggests the previous three descriptions of Bezalel—skill, intelligence, and knowledge—are all typically translated correctly, but *melakah* is not, because it also conveys the sense of keeping a business operating.[66] In other words, Bezalel was given the gift of leadership and business sense. It is the gained knowledge that comes through experience and confidence. He is able to lead his team of artisans, coordinating their tasks while ensuring excellence in quality.

Craftsmanship is talent formed through mastering one's trade. Experiential training develops the artisan's ability to do something with the materials that someone without the ability could not accomplish, harnessing both creative imagination and professional workmanship. Bezalel can envision the final product prior to its actual creation. Just as God demanded that only the best materials be used in the development of the tabernacle, the fashioning together of those materials had to be done in a way only an authoritative expert in a trade could do, without error or haste.[67] The tabernacle is not

---

[65] Levine, "The Descriptive Tabernacle Texts," 308.
[66] Otto, *Bezalel*, 58.
[67] Durham, *Exodus*, 411.

created by happenstance.[68]

For the technical artist, the business sense suggests the ability to balance budgets, schedule volunteers, recruit, and invest in the team and church mission. Craftsmanship is the difference between controlling basic fader volumes of sound inputs, and understanding the audio dynamics of each instrument, processing through various EQ, compression, and effects, and compensating for room and loudspeaker dynamics in order to create the optimal congregational experience. The *CTS Prep* offers a tangible example in regards to loudspeakers placement:

> How the loudspeakers are set in a space depends on the need and purpose. . . . You're going to want to think in terms of the experience, not the gear. Do you want to draw everyone's attention to a central point, or do you want the listening experience to seem the same throughout the space?[69]

In this example, craftsmanship dictates the difference between simply using speakers to project the sound and being aware of the experience that loudspeaker placement creates, i.e., using point-source or distributed audio, where one would draw the attention to a focal point like the stage, and the other would surround the listener in all directions. *Melakah* considers the experience of the congregation and not the ego of the artist.

TEACHING CAPACITY

The one significant difference between the two construction-commissioning narratives of Exodus 31 and 35 is that only in the second are Bezalel and Oholiab additionally endowed with the gift of teaching. This quality allows them to serve as the key leaders building up the team for both completion of the divine task as well as its continued upkeep. This demonstrates that their skills were not solely for themselves for a one-time project but were to be passed on to others and passed down from generation to generation in order to create a societal culture of skilled artisans. The gift of teaching must not be overlooked. The ability and command to teach denotes that the gifts they were given had special meaning. The skills were important enough for others to know and master. This allows the whole community to receive the skills necessary to fulfill God's calling in themselves. The Exodus 35 construction narrative specifically points out that Bezalel and Oholiab are each

---

[68] For example, craftsmanship is the difference between a Ferrari and a Honda. Both cars perform the same task and can get a person from A-to-B, and the builders of those cars fulfilled all three characteristics of skill, intelligence, and knowledge. It is the personal hand-crafting, optimization in performance, brand development and management, and quality of materials, in every aspect, that makes the Ferrari superior.

[69] AVIXA, *CTS Prep*, 137.

able to perform every task as well as teach others to do so; they become masters in a system of apprenticeship, putting together a body of professional interns so that their skills would transcend themselves. Technical skill, knowledge, ability, understanding, and craftsmanship are to be shared and passed down. Their craft was to be an act of gifting for the greater good of the Israelites. While Bezalel and Oholiab are presented here as the leaders filled with the knowledge and skills, by adding the ability to teach, their art becomes a continual work in progress with each rendition building on the previous. Craftsman inspiring craft; craft inspiring craftsman.

Their ability to teaching is directly related to the vocational aspect of their task. Even though artists today commonly think it is their art that is to support them, teaching is often the financial vehicle that allows them to perform their art and pass it on to future generations. While many artists believe that they must "resort" to teaching when they cannot support themselves from their arts, the tabernacle narrative suggests that teaching is equally important.[70] Teaching the craft becomes a countermeasure against pride and idolatry. It means the technical artist must humble himself or herself to share all his or her skills in order for them to be mimicked and improved upon. It is not about the artist demonstrating how proficient he or she is, but how proficient he or she can make those who serve alongside. Being a teacher means being a leader and mentor to others who will be called to build the "tabernacles" of future generations.

The text does not place any limits on the scope of the instruction. Being artisans "filled with the Spirit of God"—endowed with the skill of imparting their knowledge—meant that teaching the truths about God was an essential aspect of the mentorship of the tabernacle workers. Therefore, in addition to the physical and mental aptitudes being shared, so too would the knowledge of the Spirit. In this way, the pedagogy is both visible and spiritual in nature. When the technical artists share the Spirit of God, the relationship changes from teacher-student to mentor-mentee. This makes the role of the technical artist evangelistic in nature. Encouragement and guidance give the physical acts of creation their greater significance because the workers are imparted with understanding of the reason and purpose behind their creation. When workers know the objective behind the directions they are given, they become part of the creation itself. In this way, even though Bezalel was the chief craftsman, by serving alongside the craftspeople entrusted to him, his actions become pastoral in nature. By equipping the artisans with the knowledge and work of the Spirit, they become empowered to work with a directed purpose that makes disciples who create disciples.

---

[70] Veith, *Gift of Art*, 26.

BEZALEL: AN OLD TESTAMENT PARALLEL TO CHRIST?

That Bezalel's handiwork in constructing the dwelling place of God on earth was done through both creative works and a filling of the Spirit of God should not be overlooked. "Bezalel was called out, filled with the Holy Spirit, and then he translated that inner experience into multi-sensory space others could experience."[71] The experience was to worship God and seek redemption from him. Much of Bezalel's work can be viewed as a parallel to Christ and the redemptive relationship between God and his people. The actions of Bezalel, as a Christ-figure in the Old Testament, live up to his namesake, being the shadow and image of God.

It has been suggested that Bezalel's tabernacle is not merely a construction project but instead is a narrative image of the coming Christ: Bezalel is a Christophany.[72] That is, Christ appears to the people in the OT prior to his incarnation. This is indeed a risky pronouncement, and to make that claim treats the Bible the wrongly. Just because we can see commonalities in actions between two past biblical characters does not mean the two are related in person; correlation does not necessitate causation. Even though Christ—as God—would have had the ability to expose himself in the OT narrative, a better understanding is to view Bezalel as a model for becoming Christ-like. Christians would mostly agree that their role on earth is to become further conformed to the image of Christ (Rom 8:29). Thus, when a Bible character possesses similar qualities to a particular contemporary demographic—church technical artists in this case—that character can be used as a model for perfecting Christ-like characteristics in a tangible, real-world sense.

Every aspect of the tabernacle narrative says something about the character of God. "God is in the details."[73]: every detail declares something about God and his redemptive purpose. The tabernacle brings God into the midst of his people, offers them a place to worship, and an opportunity to seek reconciliation. Accordingly, there are six major concepts demonstrated within the person, works, and purpose of Bezalel that can be viewed as paralleling the characteristics of Christ for which contemporary technical artists can mimic in the midst of performing their practice: (1) the indwelling of the Holy Spirit; (2) the construction of God's sanctuary and dwelling place; (3) a royal tent fit for a king; (4) deliverance and redemption; (5) the stirring and building of souls; and (6) holy proximity.

---

[71] Otto, *Bezalel*, 140.
[72] Fesko, *Christ and the Desert Tabernacle*, 13; Otto, *Bezalel*, 89.
[73] Otto, *Bezalel*, 13.

THE INDWELLING OF THE SPIRIT

The artistic qualities of the Spirit present in Bezalel, Oholiab, and the tabernacle artisans, are noted above, yet there is another significant fact in regards to the Spirit: the gifts Bezalel possessed came from the Third Person of the Trinity, the Holy Spirit (*ruah Elohim*), literally "the Spirit of God," not just from God. Bezalel is a new creation—his abilities perfected—capable of serving God in a way that was previously not possible. "The Spirit of God would *show itself* in the extraordinary skill" of Bezalel and Oholiab, suggests Duane Garrett.[74] Likewise, the only time in Scripture where the treasury is overflowing with human generosity is also the place where the Spirit fills the body of artisans who are serving God's calling. In this way, God is in their actions by indwelling them.

In the Babylonian Talmud, Tractate Berakoth, Bezalel is spoken of as the only one "suitable" for the task because he, Bezalel, is the only one who "knew how to combine the letters by which the heavens and earth were created" (Folio 55a). It is noteworthy that an activity that would customarily be reserved for YHWH is attributed to Bezalel. The Talmud description is nearly identical to the exodus narrative, with the exception of adding this phase acknowledging he is the only one who understands how the universe was created as well as confirming him as a leader over the community. The passage concludes by comparing Bezalel to the Lord as both possessing wisdom and understanding, with further confirmation of YHWH filling Bezalel with his Spirit. This should not be overlooked. It demonstrates that Bezalel was at minimum held in high esteem by the Jews and at maximum viewed with equal powers as YHWH. Bezalel is granted authority to utilize the wisdom and knowledge as leader and creator. Berakoth 55 appears to imply that Bezalel possesses the same powers associated with the Messiah due to the Spirit within him.

Most contemporary scholars, however, hold a different view of the Spirit's indwelling Bezalel. Most suggest that Bezalel's filling with the Spirit is not his being a new covenant convert 1400 years early. It was only to help him with the task at hand. The Spirit becomes a sign of what is to come and for what will be accomplished through a Spirit-filled artisan who responds to the calling. Whether the Spirit remained with him after completing the task is debated.[75] Yet, having the Spirit from *elohim* means that Bezalel possesses the same eternal Spirit that was present at creation, and is the same *elohim* that

---

[74] Garrett, *Exodus*, Exod 31:1–11, emphasis mine.

[75] Janzen (*Exodus*, 368) suggests the filling is different because being filled with the Spirit "of God" is not as personal as being filled with the Spirit "of Yahweh." The impersonal *elohim* is used to describe God rather than the personal name "Yahweh." Though true in a general sense of the usage, Janzen's conclusion is problematic because Yahweh simply denotes the personal name and says nothing about the relationship with the believer.

blesses Jacob in Genesis 32:28.

Lastly, the indwelling of the Spirit is a "charismatic divine power" that directly infuses the person.[76] The Spirit that moves Bezalel and the tabernacle artisans is not God the Father but God the Spirit. In the same way, Jesus's ministry—like Bezalel's—only began once the Spirit came upon him (Matt 3:16, Luke 3:22). Thus, the term "indwelling" can be viewed as the most appropriate understanding of being "filled with the Spirit," because it is what dwells within that creates the outward actions.

GOD'S SANCTUARY AND DWELLING PLACE: THE TENT OF MEETING

Worship consists of two parts, the God who is invisible and the surrounding space that is visible. YHWH redefines the distance between him and humanity through a sacred dwelling place among his people so they may worship him. Being portable allows the tent to be with his people always as a defined space with distinct ritual, no matter where the journey to the promised land would lead. Even while the Israelites possessed no land of their own, they could remain in the presence of their God. In Exodus 29:46 God declares: "They shall know that I am the Lord their God, who brought them out of the land of Egypt *that I might dwell among them*" (emphasis mine). This is the theological goal of both the construction narrative and the entire book of Exodus.[77] The tabernacle was not built and then God invited into it, but instead God commanded that it be built so that he could voluntarily dwell among his people; God served as the architect, not man.

God chose to *shakan* (dwell/tabernacle) with the Israelites, leading his people to their "highest human goal" of dwelling in God's presence. No situation they would face along the way to the promised land would keep God from being among them. In this way, "Bezalel had to become a Tabernacle. . . . He is the first person in the Bible to let the Word become flesh."[78] YHWH's presence in the tabernacle suggests God's active and continual actions through Bezalel. The words used (*mishkan* and *'ohel mo'ed*), meaning "dwelling place," "tent of assembly," or "tent of encounter," denote a place where the Divine and human meet and interact. Though the terms are specific and not interchangeable, they equally express the cult of YHWH and the practices that will be involved alongside his immanence and presence. When writing about the construction, the author uses *mishkan*, but in the passages involving purpose and service for the Lord, there is an abrupt change to *'ohel*

---

[76] Dozeman, *Exodus*, 676.
[77] Klein "Back to the Future," 271.
[78] Otto, *Bezalel*, 20–22.

mo'ed. This signifies that the Bezalel narrative is presenting a cultic function of the tabernacle. The tabernacle narrative thus takes on the idea of an "event" rather than a "structure." The dwelling place and encountering of God come together when the form of the tabernacle couples with the function of the tabernacle. Even in the wilderness, while God is working on creating a permanent home for the Israelites, he produces a home for himself among them, through a construction project like none other.

Whereas God is omnipresent, the main problem lies in humanity's need to confine God to a particular place in order to devote themselves entirely to him. Thus, God concedes to be with his people in a way they can respond, given their limitations. God's dwelling place is for the benefit of the people, not the benefit of God. Regardless of whether the people can fathom YHWH outside of the sacred space among them, the tabernacle becomes the place where God can invite his people in so that he can reside among them.

The phrase "tent of meeting"[79] is employed where eternal salvation meets God's people daily in worship. The tabernacle can be viewed as a multi-purpose tent where the word-choice is dictated by description, usage, state of holiness, and presence of God. Yet, even with these various portrayals, there is only one purpose: God's dwelling among his people, as demonstrated through the narrative's climax in Exodus 40:34 with the glory of the Lord filling the tabernacle, the pillar of the Divine Shekinah denoting the precise placement of the ark and mercy seat, and thus the exact location of God within their midst. The final mention of "tent of meeting" occurs when Bezalel's filling by the Spirit is used to invoke his craftsmanship. Thus, God's purpose is fulfilled in Bezalel's actions. Outside of the garden, God shows no desire to dwell among his people prior to the commandment to build. In this way, it can be said that Bezalel is not simply God's human architect, but the actual medium for which God is able to dwell with his people, similar to Christ.

Bezalel, the OT technical artist, gives a physical presence to the God they could not see, offering the people of Israel a way to respond to YHWH's saving grace. At LeadLab Orlando 2017, Brent Allan, lighting and staging director at Northland Church in Longwood, FL, explained that his church was designed to create an experiential involvement: "wherever you are sitting, it is all around you. It is an immersive experience no matter where or who you are."[80] Every architectural detail from the ceiling beams to side walls is

---

[79] The Pentateuch mentions two tents: (1) the tabernacle; and (2) the tent of meeting outside the camp. These two tents—even though spoken of individually in the Bible—have scholars conflicted in regards to their identity. Some theologians treat them as presented in the narrative and wholly separate, while others view them as redactive revelations of the same tent. It is possible that P developed one from the other in order to bring prominence to a central sacred space of worship. Because of the various literary uses for the tabernacle, I tend to agree with the redactive conclusion.

[80] Brent Allan, *LeadLab Orlando*, Longwood, FL, June 12, 2017.

specifically lit to combine the traditional church details with a contemporary multi-projection video involvement. Visuals combine with the audio to bring the congregation "into" the worship experience. In this way, the technical artists at Northland Church become modern Bezalels paving the way for earthly ritual for the reconciliation and atonement of sins within a created contemporary tabernacle representation.

A ROYAL TENT FIT FOR A KING

Where the temple in Jerusalem is presented as Solomon's royal enterprise, the tabernacle is God's commandment. Yet it is no ordinary tent; it is a tent formed from commonly associated royal materials, fit for YHWH, the narrative's Divine King.[81] Over one ton of gold, four tons of silver, and nearly three tons of gems cut in the form of signet rings (a common ANE practice) are used in the project. Bezalel's design incorporates items customarily used for trade and commerce and repurposes them for aesthetic objectives. God's presence makes the materials valuable. Yahweh fills the kingship position normally filled by earthly kings in other ANE writings. The author incorporates a common "royal ideology" to present "divine authority."[82] Because the call to collect contributions came before the details about what to build, the passage signifies a divine speech: it serves as a commissioning project. What God commissions through the tabernacle is himself. The people, in faith, trust the command of the Divine King.

There is a royal historical connection from Bezalel through his father, grandfather, and the tribe of Judah that leads to David, Solomon, and eventually Jesus. Bezalel is, therefore, part of the blessings of Moses. The items that Bezalel, from the tribe of Judah, created would eventually be placed in

---

[81] Strong, *Tabernacle of Israel*, 77–96. Strong gives the most definitive breakdown of the meanings of colors, materials, worship elements, and furnishings used in the tabernacle construction and how they relate to common royal and divine themes. His findings from 1888 are still the standard interpretation used by most contemporary scholars. These can be summed up as such. Colors: black (shade), white (purity), blue (sky), purple (royalty), crimson (blood), yellow (sun). Materials: wood (support), copper (durability), silver (clearness), gold (value), linen (cleanliness), wool (warmth), goat's hair (compactness), ram's skin (protection), fur (softness), rope (strength), gems (hardness). Worship elements: water (regeneration), fire (zeal), flesh (substance), fat (choice), blood (life), flour (vigor), oil (richness), wine (cheerfulness), salt (wholesomeness), spice (acceptability). Sanctuary furnishings: court (special ministry), laver (piety), altar (consecration), holy place (functional priesthood), candelabrum (intelligence), showbread table (conscientiousness), incense altar (prayer), most holy place (representative high priesthood), mercy seat with cherubim (deity), Shekinah (general theophany), cloud (outward guidance), mercy seat (grace), Urim and Thummim (inward guidance), tables of the law (ethics).

[82] George (*Israel's Tabernacle*, 165) suggests that the common examples of royal curtains and yarns parallel divine and royal iconography of the ancient Near East. In turn, these imply a royal status for Yahweh. The list of materials found throughout the narrative are the same materials that would have been used in ANE royal palaces and temples. The tent was being built for temporary purposes yet with items that signify a permanent kingship.

the land of Judah. To both David and Solomon, Bezalel could feasibly have been viewed as a family hero and placed at a seat of importance. This perhaps explains why the Chronicler cites Bezalel alongside Solomon in 2 Chronicles 1:5. His role in constructing the people's temporary worship center served as a model for erecting Solomon's permanent Divine dwelling. In addition to being royalty, David and Solomon are likewise artisans—poets and musicians—whose artistic works would themselves go on to be significant worship and instructive writings in both Jewish and Christian Scripture. Even though Bezalel is not an actual king, he possesses many of the same qualities as other biblical kings. This is important to the overall argument because it suggests that the work performed is also authoritative. Likewise, technical artists who create artistically for use in God's sanctuary ought to use materials and techniques in a way that are not chosen by happenstance but fit for a king. For the Christian that King is the Divine one in the person of Jesus. What they do, they do as if directed to and for Christ. "Just enough" is not enough: quality matters.

DELIVERANCE AND REDEMPTION

"First salvation, then worship."[83] God demonstrated his love for his people by delivering them from Egypt, and then he gave them a permanent sacred space in which to worship him, even while their homes were temporary. The construction narrative is the first biblical instance of divine forgiveness, tying together the creation narrative with the reclamation of God's chosen from Egypt. While the people are expected to obey the Lord, he demonstrates divine forgiveness by selecting Bezalel to build his dwelling space, even after the people's disobedience in worshiping the golden calf. In this way, art can be symbolic: "What is symbolized . . . is the very gospel itself."[84] The tabernacle ritual and furnishings create a simultaneous physical representation of separation and inclusion. Bezalel performs the redemptive work by mediating God's coming into the presence of humanity and allows a space for the Israelites to worship and offer sacrifice to their Lord. Bezalel serves as the model for the renewal of humanity by creating a sacred space for the inclusion of God's people in the worship of him. In this way, technical artists are able to create pleasing art through sound, light, and visuals that drawn the people of God toward the dwelling place of God in order for him to commune with them.

The tabernacle, designed by God, with supplies provided by God, through artisans selected by God, demonstrates God's desire to redeem his

---

83 Peterson, "The Pastor," 95.
84 Veith, *State of the Arts*, 123.

people. The narrative at chapter 35 transitions from God's actions and commandment to a lifetime of participation in redemption and revelation, serving as the physical expression of the restoration of the relationship between the Israelites and God. God recognizes that his people needed a permanent reminder of his mercy. Thus, the entire chiastic Bezalel narrative can be viewed as a creation, fall, and recreation-redemption motif. The construction can only happen after Israel's rebellion. Until then, the tabernacle is only blueprints.

In the tabernacle ritual, blood is poured out. Just as priests would use animal blood at the altar to atone for the sins of the Israelites, it can be suggested that Bezalel is an OT messianic symbol who creates the sacrificial space for the blood to be shed for God so that God and humanity can have eternal fellowship with each other. There is completeness in Bezalel's construction. Expressions of atonement remind God's people of their need to worship and serve him. Where God comes to redeem humanity, humanity's response is worship and sacrifice. The tent of meeting, where eternal salvation meets God's people daily, has similarities with Christ's election as the sacrificial Lamb, redeeming God's people with the result of turning to and worshiping him. Where tabernacle ritual has blood poured out ritualistically, Christ's sacrifice of himself meant that his blood could be a once and forever atonement. Bezalel and Oholiab create the aesthetic background for the power of sacrifice to draw people to God in the same way that modern artists create the atmosphere for presenting the reminder of Jesus's sacrifice for them.

The new beginning in salvific history takes place at the exact one-year mark from fleeing Egypt when the tabernacle is dedicated and God takes his sacred place among the people. Thomas Dozeman notes: "The revelation of covenant law and the eventual residence of God in the sanctuary transform the abstract space of the wilderness into a place of value filled with meaning."[85] Up until chapter 34, the Exodus narrative is about salvation and revelation. At chapter 35 it turns to redemption, preparing for the continuing worship of God that weaves together salvation and revelation into the fabric of normal daily life. This reflects the people's gratitude for God's deliverance from Egypt. But it is more than that, it is an act of repentance. The people demonstrate repentance through the voluntary building of God's holy sanctuary, through both giving and responding to the Spirit-filled calling. Where worshiping the golden calf became the ultimate denial of the freedom God delivered, the technical artists lead the redemptive action by completing the divine construction project. Just one year into the wilderness experience, God redeemed his people as part of the forty-year wandering. For the next thirty-nine years, the people were not alone; they had their redemptive God in their

---

[85] Dozeman, *Exodus*, 574.

midst at every stop along the way.

> Moses led people to salvation freedom; Bezalel paid scrupulous atten-
> tion to the details of that freedom embodied in a holy life. Moses
> brought down the Ten Words from Sinai; Bezalel assembled them co-
> herently in acts of offering and sacrifice. Moses and Bezalel.[86]

For the contemporary technical artist, this suggests that they do and
should work hand-in-hand with pastoral leadership. Where pastors and
teachers lead the congregation to know God, technical artists turn that
knowledge into active reminders of the work of God. Church leaders present
humanity's redemption through Christ; technical artists create tangible ways
to experience that truth.

STIRRING OF SOULS

Bezalel accomplishes his massive construction project because God
makes it internally possible. The result is a response that encompasses the
whole of one's experiential sensory capabilities; for those who enter, it is a
full-body experience of all tribes, genders, and family, through obedience to
an inner stirring from the Lord. It is a response made in faith. The tabernacle
text is at odds with other ANE building accounts because focus is placed on
the people doing the tasks rather than credit given to the manager or deity.[87]
It is God's people who receive the credit. Due to this, there is a danger in
losing the social aspect of the creating by placing focus on the individual
worker rather than the purpose and person for whom it is created, thus creating a "hero worker crafting art."[88] Yet, this is precisely the point of the
narrative. Bezalel is the picture of the perfect obedient sacrificial servant. The
text says something about the person who responds to an internal stirring by
the Spirit within. When performed as a community, the individual gifts and
abilities of the artisans contribute to the whole, and in turn, the creation joins
everyone together in recognition of the source of their abilities. In the case
of the tabernacle experience, it would bring recognition to the person from
whom the Spirit was given.

While sharing the strategic vision for the technology department at the
University of Central Florida, Don Merritt proposed that "Scale x Excellence
= Impact."[89] Exceptional and challenging tasks like that of the tabernacle—
when performed to a high standard of excellence—are able to change and

---

[86] Peterson, "The Pastor," 96.

[87] Kidwell, *Theology of Craft*, 36. Even if Moses attempts to take the credit, as Exodus 40:18 suggests, he is
never mentioned as the project manager.

[88] Ibid., 46–47.

[89] Don Merritt, Bradley Jones, Todd McMahon, and Anna Turner, "Transitioning Users to New Technology," *InfoComm 2017*, Orlando, FL, June 15, 2017.

direct lives with greater impact. Because the Spirit stirs the souls of Bezalel, Oholiab, and their fellow artisans, they are moved to serve God in a way not previously seen. Where constructing a worship center would have previously been an act of slave-work for Pharaoh, the people freely bring the best materials and craftsmanship to create the sacred space for their God because their hearts were stirred.[90] Where kings in other ANE cultures would demand forced labor, the Israelites demonstrate their reliance on the Lord through giving abundantly and serving voluntarily. In turn, the Israelites demonstrate their obedience by following the original command in full.

God does not select the materials donors personally, like he selected Bezalel and Oholiab. The offering is open to all. Likewise, there are no consequences for not giving. It is an open and free opportunity to give as the heart is stirred. The congregation nonetheless responds to the funding call in order that Bezalel can fulfill the commandment given to him. They donate to the extent that Moses calls for an end to the giving due to the abundance of donations (Exod 36:3–7).

HOLY PROXIMITY

Regardless of the physical realities of the tabernacle, its theology is tied to the holiness of Israel. There are progressions of holiness, with the power and immanence of God increasing as one moves closer to him. The laypeople reside in the outer court, priests in the holy place, and the high priest in the holy of holies. There is an inherent social structure built into the instructions, a hierarchy of authority, mimicking the value of the materials used: from bronze to silver to gold. For the whole to become holy, so must each individual part that points to God. The tabernacle itself becomes a structure of separated sacred and profane space. In this way, it can be suggested that God demands greater holiness in order to approach him. Thus, for those who create the worship designed to connect God with his people, purity and holiness must embody the created works.

Likewise, the tabernacle itself maintains a high standard of quality. Just as there is a call to become holy, there is an expectation to follow all God commands, exactly as he commands.[91] No articles created for use in the tabernacle could be profane, and the law required adherence to the Ten Commandments, avoiding idolatry or creating objects to be worshiped over

---

[90] There is a duality at work. The construction of the Israelite worship center is made possible from materials gathered from Egypt. The same materials previously used to build altars for Pharaoh are given to the Israelites so that they may create the tabernacle of the Lord.

[91] I contend that the artisans are able to use a certain amount of poetic license during the design process insofar as their creation was aligned with the holiness of God.

YHWH himself. Bezalel was commissioned to create art that built on qualities of God, including goodness, truth, and beauty. In a broader context, the passage signifies that God selects the people for his tasks and provides the means to achieve the goal, but the Spirit sanctifies them for it to be holy work.

Because modern-day church technical artists perform their tasks within the holy of holies of today, they are direct distributors of the Word serving in the priestly role through sound, visuals, lighting, and atmosphere. They have a duty themselves to be holy just as the priests would have been holy. Where the high priestly standard of purity was pronounced prominently—even to the point of wearing an engraved turban seal stating "Holy to the Lord" (Exod 28:36)—many Christians today commonly equate their standard of serving in the church in accordance with 1 Timothy 3:1–13, being above reproach, not able to fall into disgrace, and holding the truths of the faith in a clear conscience. Technical artists who mimic the role of Bezalel as outlined meet that standard. In this way, as merchants of God's characteristics through audio-visual-lighting, their lives become a reflection of how those truths are received and perceived by their respective congregations.

BEZALEL AS A CHRIST-LIKE MODEL:
THE REDEMPTIVE ARTISAN

Similar themes often used to describe the actions of Bezalel can also be attributed to Jesus. A comprehensive list of these parallels includes:

BEZALEL	JESUS
Builder in all craftsmanship	Carpenter
Created	The Creator
Built the predecessor to the temple	Tore down the temple
Possessed the Spirit of God	Possessed the Spirit of God
Built a tent	Called the "true tent"
Built the physical tabernacle	Was the tabernacle
Place blood is to be shed for atonement	Atoning sacrifice shedding his blood
Used royal materials	The Divine King of Israel
Tribe of Judah	Tribe of Judah
Line of David	Line of David
Referred to in Isaiah	Referred to in Isaiah
Tabernacle filled with glory of God	Brought the glory of God to earth
Space for worship	Person to be worshiped
Called by name from the Lord	Called Lord
Name = "image of God"	Is the Image of God / Is God

Materials came from first Passover	Redefined the Passover elements
Received offering from Moses and people	Person accepting offering
Created the first physical church	Built the church
Built replica of heaven on earth	Established heaven on earth
Mediator between God & humans	Mediator between God & humans
Gave all of himself to redemption building	Gives all of himself to redeem
Rejected from receiving credit and honor	Rejected and despised
Received blessing	Gives blessing
Served ANE role as builder for the king	Is the King
Levels of holiness in tabernacle	Is holy / Makes believers holy
Physical reality of the Word	Is the Logos
Used physical to reveal spiritual	Used physical to reveal spiritual
Built from and for all tribes	Great Commission embracing all
Took on high priestly role	The High Priest
Built sanctuary	Find sanctuary in him
Possessed knowledge	Omniscient
Teacher	Teacher
Given wisdom from the Father	Given wisdom from the Father
Had assistants from the people	Had disciples from the people
Artisan who created "for beauty and glory"	True Beauty and Glory of God
Stirred souls in service of the Lord	Stirred souls in service of God
Built mercy seat	Offers mercy
Built ark of the covenant	Established the new covenant
Sewed the veil	Tore the veil
Resting place of deliverance	The Deliverer

Assuming these similarities are valid—and accepting the common belief that the role of NT Christians is to become further conformed into the image of Christ—the conclusion can be drawn that a church technical artist who mirrors the practices of Bezalel is able to move further toward that end. At minimum, the similarities are noteworthy. At most, they demonstrate that a valid way to be conformed to the image of Christ is through technical artistry performed for and in God's sacred space. The evidence advocates latter.

WARNING: IDOLIZING THE CREATED CALF

Between the commissioning and construction narratives sits a polemic against the creation and worship of false idols: the golden calf. It is the "negative tabernacle" or anti-tabernacle. It is done without God's word of commandment due to the impatience of the Israelites. The text demonstrates the pitfalls of misdirected worship of humanly produced items. At the same time the instructions were being given to Moses, the people were deciding their own worship methods by creating and worshiping the calf. Where the calf was intended to represent Yahweh, the narrative irony demonstrates that in the act of creating something in his image, they simultaneously defamed the same God currently offering Moses the plans to dwell with them. Where God gave the tabernacle as a gift and personally provided the materials to create it, the text suggests there is both a good and bad use for the items. The same gold pillaged from the Egyptians is the same gold Aaron crafted into the golden calf and the same gold used in ornamenting the tabernacle. Kevin Bauder thus contends: "Some judgments [on created objects] are better and some are worse. Some are devastatingly bad, for worship that is not according to truth is simply idolatry."[92] In this way, it is problematic when the means and methods supersede the intended purpose.

Likewise, in terms of worship, Deve Ganga-Persad suggests:

> [The golden calf narrative] gives a picture of how fickle and wayward the heart of people can be. We have a tendency to need to worship something and when we lose our focus on who God is, we are quick to replace that position. . . . These actions taken by Aaron and the Israelites are what provide definition to idolatrous adoration. Worship and credit is attributed to an object that they had created with their hands and with material they provided.[93]

When adoration is fixed upon an object other than God, it becomes an idol. For the technical artist, this appears as a real concern. As discovered in the documentary analysis—with few resources dedicated to the theological understanding of the craft and over three-fourths of the literature aimed at gear and technique—focus can become be easily misplaced, as if the tools themselves generate the worship rather than the internal work of God through the artist and congregation. While the text does express that tangible skill and craftsmanship are important aspects for becoming a biblically in-formed technical artist, and like any artist, tools of the trade are important to quality, when the artist places the "things" over the innate ability gifted by God—and enhancing those skills—the heart becomes misdirected.

> Where your heart is, there your technology will be also. . . . It shows whether your heart is oriented toward God and toward finding true joy and satisfaction in him, or whether you are attempting to find counterfeit joy in the things he forbids.[94]

The tabernacle narrative places God as the DNA—filling it with his glory—not any one particular element within the sacred space.

---

[92] Bauder, "Why Pastors Should Be Learned in Worship and Music," 14.

[93] Ganga-Persad, *A Theology of Image*, 5–6.

[94] Tim Challies, "Letter to Teens Unboxing Their First Smartphone," Desiring God, May 26, 2015, http://desiringgod.com/articles/letter-to-a-teen-unboxing-their-first-smartphone.

The text presents more than a warning against idols, but also the act of worshiping the process. The tabernacle was a continual "work in progress." Even upon completion, the tabernacle required ongoing upkeep and operation: processes and procedures defined the role of the high priest—and presumably Bezalel and his fellow artisans. Similarly, in the modern church, technical artists follow a producer's script or "order of service" that dictates every detail from camera shot to audio highlight. Post-service practices involve repairing and maintenance of the equipment along with research of the newest manufacturer tools and products. When these procedures become the main focus, idolization of the creation supersedes the worship of Christ, which is at odds the findings in this chapter. While emphasis on the art form can satisfy the requirements of beauty and function, it can fall short of conveying God's transcendence to the congregation.

Because the technical arts harness cultural norms, they are easily susceptible to falling into societal idolatry through the desire to people please, taking the place of God himself. Schultze, Chuang, and Redman discuss: "To do technology and worship well, you have to fit the technology to the worship, rather than the worship to the technology. Otherwise, the worship can become distorted."[95] It must be used "within the context of worship, not use worship to celebrate technology."[96] The presented technological language needs to match the theological exposition. Authenticity in art and mechanical reproduction only happen when the presence of the intended purpose is the prerequisite, built upon its historical tradition, and aimed at connecting that message in a clear and understanding way to the current congregational loci. In this way, the message of the creation—namely worship of God and God's wishes in gifting the tabernacle to the people—must dictate the wisest technological methods for achieving that goal. "The wrong place to start is with the right equipment."[97] Theology must inform practice. Brad Herring an interesting question:

> Think about how Jesus taught. Can you imagine him on a stage while 30 moving lights swept around Him? Does His teaching method strike you as someone who would have produced dance video productions? While no one has the definitive answers obviously, I'd like to say that I think not. The Gospel is strong and it stands on its own.[98]

While Herring is correct in two aspects—no one has a definite answer and the gospel can stand on its own—his conclusion misses the exegetical point. It is not about whether sweeping lights would be used, but how. They would be used to highlight the person of the gospel. Jesus used every tangible method for sharing the gospel possible. How was he able to preach to

---

[95] Schultze, Chuang, and Redman, "Worship, Technology, and the Church," 97.
[96] Schultze, High-Tech Worship?, 63.
[97] Huyser-Honig, "Digital Storytelling," no pagination.
[98] Herring, Sound, Lighting and Video, xv.

thousands? Because the natural acoustical design of the natural scenery; Christ himself was a technical artist in presentation method. In the same way, Lee Bloch condemns the conversations had with tech leaders who flaunt the "coolness" factor in the way they programmed lighting instead noting how the lights allowed the congregation to either view the presentation better or served as a compliment to the message presented via lyrics or pastoral message. Yet, again, finding beauty ("coolness") in the product is not against the text—for indeed rich aesthetics is a core aim. Idolization of the created calf comes when the aesthetic qualities are the aim over the object that the product reveals. There is no need to improve on the gospel, but it does need to be pointed to; that is the role God has gifted and entrusted technical artists to fill.

Technology for the sake of technology is contradictory to the attributes of wisdom and knowledge presented of Bezalel. Kevin Bergeson notes: "Using a clip of leaping whales may be a powerful way to illustrate a creation story, but what happens when people leave worship remembering only the whales and not the God who created the whales?"[99] Likewise, Bernard Reymond warns of Contemporary Christian Music's "veneering of rock melodies with Christian words as if it sufficed to say 'Jesus' in a song instead of 'drugs' or 'violence' in order to transform it into a Christian song!"[100] This is important in church tech practices because the skills used so closely parallel those of the secular world. The liturgical line between entertainment and worship is a real one. For example, oversized screens with Image Magnification (IMAG) prevalent in large sanctuaries are often used for more than displaying song lyrics; they create a show in themselves. Attendees often only feet away from the pastor will still instead view the service "virtually," being diverted from the preaching directly in front of them, "watching," rather than staying engaged in the live message.

Technology, therefore, "can be good or bad, suitable or distracting; the word can be vitally proclaimed or recited in a meaningless, ritual fashion."[101] This suggests that the technical artist ought to be more than a technical expert, but also a liturgical planner, knowing how the technology can be harnessed for the creation of sacred space by utilizing proper means and methods without idolizing the product. Idolization of the process appears to be a common roadblock from keeping the gospel in view. "[It] comes down to being seduced by entertainment or utilizing entertainment to reach people for Christ."[102] The church, and its technical artists, "must decide what it cares

---

99 Bergeson, "Sanctuary as Cinema?" 303.

100 Reymond, "Music and Practical Theology," 88. Style is not the concern of the text, but substance. While Reymond's point about not simply exchanging drugs for Christ is valid, when musical styles are imbedded with the gospel of Christ, the medium becomes the message.

101 Viladesau, *Theology and the Arts*, 57.

102 James Tippins, "Music and Drama: Entertainment?" *Worship Leader Magazine*, Accessed April 15, 2015, https://worshipleader.com/culture/music-and-drama-entertainment/.

about the most: The product or the people. Churches should remember that people are the product."[103] For the technical artist, the golden calf narrative serves as a warning against making the product the purpose rather than using the product as a means for God to connect to his people.

CONCLUSION: THE TECHNICAL ARTIST AS A MODERN BEZALEL

From a doctrinal standpoint, the Bezalel and Oholiab narrative suggests that artistic works can be consecrated for use in the service of religious practice. The works do not dominate over religion but offer a supplement to worship practice. The biblical text indicates that technical artists require the five physical qualities: ability, intelligence, knowledge, craftsmanship, and teaching, with the sixth supernatural quality—the Spirit—making each useful for producing worship of God. In doing so, the works made create an object of greater power than that of the artisan himself or herself, and one that mirrors a life aimed at becoming further conformed into the image of Christ. Technical artists serve the same roles as Bezalel and Oholiab as ministers of the redemptive qualities of the gospel when their art is performed in response to the internal calling of the Spirit.

---

[103] Stephen Beasley, "Backtalk: People vs. the Product," *Church Production Magazine*, Accessed June 7, 2015, https://www.churchproduction.com/ministry/people-vs.-the-product/.

❧ 4 ❧

HEBREWS 2:12–13: JESUS AND THE TECHNICAL ARTISTS' PARALLEL ROLES AS MEDIATORS OF CHURCH WORSHIP

2:12a The Son says to the Father,
    "I will proclaim your name
      (of who you are
       and what you have done
        through me) [1]
    to my brothers and sisters.
2:12b I will sing a hymn of praise to you
     in the midst
      of the Church congregation."
2:13a And again Christ says,
    "I will go
      put my trust in God,
       the Father."
2:13b And again Jesus says,
    "Behold, here I am,
      I
     and the children
      with whom God has given to me."

---

[1] Continued from the declarations established in Hebrews 2:10-11.

INTRODUCTION

Hebrews 2:12–13 is a central passage for developing a theological under-standing of the technical arts, holding significant implications for modern church worship practices. The verses firmly establish the gospel message de-scribed in terms of revelation and response: God reveals himself through the deity and humanity of Christ, and believers respond in song alongside Christ the Mediator. Most of the scholarly literature on this verse focuses on the surrounding context, overlooking the significant fact that Christ, in his ex-alted state, is leading his people in worship of the Father from within the church congregation. Christ-exalted is still present. Hebrews emphasizes Christ's completed work on the cross yet does so in terms of his continuing earthly ministry. Just because Christ-exalted is seated on the throne at the right hand of the Father does not mean he is inactive in his earthly work. Hebrews professes that Christ "always lives" (7:25) to make intercession for those who seek the Father through him. Just as Jesus facilitates the reconcil-iatory relationship between the Father and humanity through being the Me-diating, Singing Savior, the verse serves as a metaphor for church technical artists who mediate the worship experience between the stage and congrega-tion, and thus between God and the church.

This chapter builds upon the foundation laid by John Paul Heil and Ron Man in their respective works *Worship in the Letter to the Hebrews* (2011) and *Proclamation and Praise* (2007). Hebrews is more than a theological treatise, or even pastoral sermon, as is commonly understood: it is itself a worship man-ual. It instructs church congregations—both ancient and modern—how en-durance through living out their faith is a tangible act of worship. This is done through the act of a worshiping Christ among his brethren-children. In turn, the author's Christological example mirrors the work performed by the mod-ern church technical artist who serves as a mediator between the stage and congregation. The view taken here is that in Hebrews 2:12–13 the church, seen as God's people, can mediate the gospel to the congregation through worship practices and technical productions.

EXAMINATION OF THE IMMEDIATE PASSAGE

The immediate issue of concern for the author is apostasy. He encourages his audience to remain focused on God because of the living example of Christ rather than returning to their former customs and deliberately aban-doning Christ as Messiah. The author's appeal is not merely a push to endure but a caution against the perils of turning away from the faith, as demonstrated through the five warning passages, all of which speak to the misguided thoughts of the congregation. The author is not concerned with assurance of salvation but the product of an active faith lived out in the promises of God, urging his Jewish-Christian congregation to leave the synagogue and fully identify publicly with the church. In fact, the entirety of Hebrews revolves around God's relationship with his children, even if the immediate context is not apparent in this goal.[2] The book focuses on a relationship between believers, Christ, and the Father, enabled through what Christ has done as well as who he is in his continuing ministry, that by God's grace he would taste death for everyone, being "perfected" as the "pioneer" of the salvation of his children and brethren. Christ confirms the relationship established by calling those who believe his brothers and sisters, further serving as their helper and mediator. The audience is to think not only of what Christ has done; they are also to recognize the future he promises through obedience and keeping one's eyes fixed on the Son. Without Christ's active and present participation, the call for immediate obedience falls short. In this way, the epistle is to be read in light of the urgent relational pronouncement between the Father, Christ, and his children. In doing so, Hebrews becomes a manual for ecclesial worship.

The surrounding passage of Hebrews 2:5–18 completes the description of the eternal and exalted Son started in chapter 1 by displaying how Christ's incarnation and suffering are the means to his becoming the sufficient Savior through establishing the soteriological significance of the Son's death. In turn, solidarity between the Son and believers is established. Christ is presented in his humanity, yet still higher than angels, wherein God became the perfect mediator through human suffering. Suffering and subjection is part of God's perfect plan; it is a Christological claim that in his suffering, Christ shared in the humanity of those he came to save. In this way, both believers and Christ have one common Father. Therefore, Jesus is not ashamed to dwell, suffer, and call them; Christ shares an intimate relationship with his brethren-children because of his humanity. The word prepo in 2:10 translates to "it was fitting;" the only suitable way God could reconcile sin was through the incarnation and death of the perfect and superior Son. To Christ, his sacrificial payment is not humiliation, precisely because he finds solidarity among his brethren fitting. This is significant in validating Christ's dual nature as fully man and fully God and, therefore, Christ's role in the church and worship.[3]

---

[2] Attridge, *Epistle to the Hebrews*, 103–8. The author's immediate instruction is directed at the congregation and is offered as an a fortiori metaphor for God's instruction to his church. The author is arguing that since God has reconciled himself with humanity through the personal, saving work of Christ, how much more then should a faithful, lived-out response be in return to God?

[3] God reconciles his people to himself through a human state. Therefore, Christ's earthly connection to the Heavenly Father is through the example of earthly worship and singing.

Scholars commonly cite the OT references in Hebrews 2:12–13 as proof texts to demonstrate the Son's confession of his relationship with his people as viewed through his deity, humanity, and redemptive work. The pastor-author moves across chronological and historical distinctions by taking the words of David and Isaiah and making them the present proclamation of Christ. Hebrews 2:12 is a near mirror image of the author's LXX version of Psalm 22:22—LXX Psalm 21:23. The only difference is a replacement of the future indicative form of *diēgeomai* (I will tell/describe) with *apangellō* (I will tell/proclaim). Likewise, 2:13a is only slightly altered from the LXX in word order of Isaiah 8:17, moving *esomai* forward and adding an emphatic *ego* to open the phrase, giving Christ a position of power and self-declaration of his chosen action. Hebrews 2:13b finds full agreement with the LXX in Isaiah 8:18. By placing the words of these earlier prophets on the lips of Christ, Jesus becomes the fulfillment of the prophetic voice. Extracting only the changes between the two texts, the author emphasizes an authoritative promised action of Christ, saying, "I will proclaim!" The focus changes from man praising God to Jesus himself proclaiming solidarity with his position as brother among the brethren.

A chiastic structure of vv.12–13 is in play.[4] The two declarations of "I will proclaim" (2:12a) and "behold" (2:13b) frame the two actions of Christ singing praise (2:12b) and placing trust (2:13a) in God the Father. This structure emphasizes Christ's purpose among his brothers and children through his singing within the assembly and trust in the Father's ultimate plan. Christ becomes the living example of what the author suggests is the proper response to the gospel action presented in 2:10–11, namely worship of and to the Father through Christ. Christ's proclamation establishes him as the Singing Savior, the promised messianic Worship Leader. He is the true worshiper, leading the congregation and offering an example of what ecclesial worship ought to mirror: praise and trust.

Discussing this in artistic terms, Craig Koester describes the exordium starting in 2:9 as an artist's painting.[5] The author of Hebrews takes on the creative quality of presenting the death and saving work of Christ from various dimensions, offering several perspectives of imagery and possibilities and creating interplay between them in order to complete the portrait of the work of Christ. The section of 2:10–18 is framed by Jesus suffering in v.10 and v.18 and then offers four "portraits" to support it: (1) what is fitting for God (vv.10–11a); (2) the bonds of brotherhood (vv.11b–13); (3) conflict and victory (vv.14–16); and (4) high priestly atonement (vv.17–18). In this way, Christ's suffering defines his saving, responsive relationship with humanity.

---

[4] deSilva , "Responding to God's Word," 115–17; Guthrie, "Temples of the Spirit," 951.

[5] Koester, *Hebrews*, 233–35.

Likewise, John Paul Heil speaks of the author's work in "levels,"[6] similar to how a painter layers elements of light and shade, a musician incorporates dissonance and resonance, or how a technical artist "layers" individual instruments in a sound mix to both "bury" and "spotlight" as the song progresses through verses, choruses, and bridges. At the highest level incorporating the epistle in its entirety, the exhorted are to be faithful to the grace of God granted through the Son. The middle level encompasses 2:10–18 encouraging the church that those being tested can be helped by Christ. At the lowest level, those tested will have sins atoned for. Though Koester and Heil may have been writing metaphorically in order to explain the author's intentions, they both picked up on an important and much-overlooked aspect of the immediate context: in this passage Christ is an artist. He is worship leader and worshiper. He leads his people to him not simply through his saving work, but he joins them in their midst through song and artistic connection with the Father through himself. The chiastic use of the three OT references become a "song" denoting trust in God because of Christ's relational presence. Therefore, the text suggests that solidarity with the Son serves the purpose of focusing on God. When Christ speaks directly to his children, he invites them to participate in his same experience, joining together as members of the family of God.

As a worshiping congregation—seen through the abandoning of their standard meeting practices (Heb 10:23–25)—the audience is instructed to engage in worship practice communally. The author writes from the position that his congregation is familiar with Christian liturgy in terms of both OT synagogue and NT house church practices, with the OT invoked in order to express a particular truth about the present situation. The appeal for liturgically structured worship practice makes the text stand firmly in the mainstream of the intersection between Judaism and early Christianity, where new habits are being formed yet old habits are equally tempting in the face of persecution. The congregation had spiritually developed enough to be persecuted for it, which suggests their worship practices similarly developed.[7]

In this way, worship is the primary concern of Hebrews from start to finish. The letter is intended to be publicly and orally performed as an act of worship to a worshiping congregation. Verbal clues and hook words suggest the letter is a midrash intended to have its points made more powerfully through oral delivery. In fact, the text is missing many of the common elements expected in a letter from this time, like an opening prayer, call for grace

---

[6] Heil, *Worship in the Letter to the Hebrews*, 174–87.

[7] Docherty (*Use of the Old Testament*, 27) suggests that scholars must be careful not to overly read the text as "good Protestants" but instead view the audience as worshipers of Christ with mixed OT to NT liturgical practices and traditions. There is a common practice of reading current practices into the early church. The first-century transition from OT to NT worship practices would have also been a time of identity seeking. Therefore, the exact rituals are less important. What is significant is that believers are to meet and develop a unique liturgy aimed at the person of Christ.

and peace, and expression of thanksgiving and blessing. Literarily, it is poetic in nature. Being a poetic worship passage framed as a sermon allows for the epistle to be understood on a more personal level; the author directly leads his congregants in knowledge of who Christ is and what he has done. For this reason, a reading of Hebrews that treats it solely as a theological and systematic treatment of the work of Christ would be flawed. The text was written to serve an instructive, pastoral purpose, offering direction for proper Christian living to the church body. With the epistle delivered to an audience of converted Jews living soundly within the Greco-Roman world, the text is to be view from both of these distinct worldviews. The epistle is not solely a pastoral exhortation but also a worship manual in which Jesus in Hebrews 2:12–13 is presented as the Singing, Messianic Savior.

THE JEWISH CONTEXT

The pastor-author wants his congregation to make the connection to the broader OT context, tying their understanding to Jewish liturgical practices while applying it firmly to their new faith in Christ. Christ was not the object of worship in the early church but was the subject of praise, hymn singing, and the content of early Christian worship. This would have developed from the way YHWH was viewed in religious practice: God was the only one worthy of direct worship, with Christ being the representative of God's actions in the world. In the NT era of Hebrews, the author demonstrates that worship is now personal; there is solidarity between worshiper and worshiped. NT believers take on the OT Levitical roles. The congregation replaces the professional Levite musicians and David's official temple choir. Whereas the Levites formerly would act on behalf of the people, Christ's role among his people invites them to join him in one concord. Thus, NT worship is an example of the reconciliation that could only happen because Christ-God became man-brother.

Since the early Christians came from a Jewish background with warnings about worshiping anyone or anything other than YHWH, the addition of Jesus as an object worthy of worship was no small task. Linking Christ to the Father had to maintain a monotheistic awareness. In this way, the pastor-author formulates his rhetorical argument by citing OT psalms and prophetic Scriptures in order to tie both the humanity and deity of Christ to OT worship practices. For example, the standard Jewish use of Psalm 22 found in Hebrews 2:12 would have normally been either cultic, or utilized only in times of illness or oppression. Israelites did not view it as messianic in the way early Christians did. Use by Jesus gives it messianic meaning, and therefore the use in Hebrews is entirely appropriate, similar to Jesus's synagogue reading of

Isaiah 61 in Luke 4:16–21, where he likewise states that he came to proclaim the good news.

Even though images of Jesus singing are virtually unknown in religious art and hymnody, it can be concluded that Christ would have regularly sung during his earthly ministry because singing was a standard part of Jewish liturgical practice. Indeed, for him not to would have been out of the norm. In the Jewish culture, music for music's sake would be unknown: music had purpose, the purpose of festival, praise, and lament, among others. With about three-quarters of the musical references in the Bible referring to singing, Israel was indeed a singing culture, and therefore so too was the early church.[8] Being led in structured oral worship by the priestly class would be second nature and expected. The use of praise songs about the person and work of Christ served as an indirect method for worship of him. In the text, the Hebrews congregation was considering returning to their previous ways. Consequently, it can be inferred that those former ways included their previous worship and singing practices as they understood them. I conclude, therefore, that the role Christ plays in Hebrews 2:12–13 is a direct parallel to the work he would have been known to perform during his earthly reign, namely leading his disciples in worship as their high priest. In modern times, it seems out of the norm because people rarely picture Jesus in this way, but Christ proclaiming and leading worship among the brethren would have been a natural role for him, and it would have been natural for his disciples to mirror him in it (i.e., Matt 26:30).

THE GRECO-ROMAN CONTEXT

God's role in Hebrews is unique in Hellenistic philosophy; the idea of a deity suffering and being shamed would be unprecedented. Worshiping someone or something in this state would be even more suspicious. The author's primary purpose for his Hellenistic audience is to establish that Christ is the worthy mediator-object of worship. The author calls it "fitting" how God has used Christ's role in humanity, even though the Greco-Roman audience would struggle to view it as such. The shame Christ endured would have been viewed as similar to the shame of being enslaved to another. Christ needed a noble death. Because dying by crucifixion was the most heinous, shameful, and disgraceful way to die, much rhetorical skill would be required to convince someone of Jewish theological and Greco-Roman societal background that his or her faith is founded in this truth.

In Greco-Roman society, how someone died had something to do with

---

[8] Begbie, *Resounding Truth*, 61.

his or her character, and thus self-sacrificial death—even if by crucifixion—could still bring honor if presented and understood in that way. The greater the sacrifice, the greater the honor and glory; the greater the righteous humiliation, the greater the honor and glory. Christ's selfless act presented him his honor. Christ was able to accomplish reconciliation through shame; rather than avoiding it, he embraced it, saving humanity rather than himself. This brands Christ's humanity as perfect and unique: contrary to conventional cultural thought, through human suffering Jesus became the spotless atoning sacrifice, bringing his children to glory. Even in his present exalted state, the divine, holy Christ is not ashamed to be reconciled to his unholy, sinful brethren-children.

Both pagans and Jews would have seen it as madness and blasphemous to place a crucified man in second order to God. This objection is overcome through a noble death that leads to salvation for all. For a community rooted in a system of honor-nobility, 2:13b reaffirms the congregation's own honor and relationship with the Father. They too are of noble birth through their faith in Christ. Though persecuted by society, to the Son they are of high esteem. Even though the future promise is still to be fulfilled, it is an honor to which they can hold fast. The author confirms the victory of the Messiah, coming full circle as the OT fulfillment. Thus, the call is to look to the Son as the agent of all things past in Scripture, present with the community, and future in glory. Therefore, because the congregation would have been fearful of persecution from both Jews and Romans, their status in Greco-Roman society was being challenged. Being under persecution, the people were not honored in society.[9] However, the author proclaims God is not ashamed, showing that their honor is secure through Christ even if not present in their current circumstances.

Jesus fills the role of the master in the master-slave relationship, granting honor that only he as Suzerain-God can offer. Care, generosity, help, protection, and security were common first-century requirements of the master within the suzerain-vassal system. In this instance, Jesus-master moves further toward becoming brother-family with the slave.[10] God serves as a Greco-Roman benefactor throughout the story of Hebrews, with Christ being the broker of the deal. It is because of the care of Jesus as mediator that the deal gets done. In obedience to his own suzerain-master, he carries out his duties for his vassals. In this way, Christ is both suzerain and vassal: vassal of the

---

[9] Kidd (*With One Voice*, 124) suggests that the looming—and possibly present—Jewish War (66–70 CE) would have added to their strife. The congregation would be living between two cultures and disliked by both due to the influence of the other. I propose this possible reality serves as a strong argument for the congregation's desire to flee the faith in order to find stability on one side or the other.

[10] The term "brother" in Greco-Roman nomenclature extended beyond the family nucleus to mean "a close connection" that could also apply to friends, Israel, and religious affiliation. This can be viewed in the same way as is still common today for believers to call each other "brothers" or "sisters" in Christ even without sharing any direct biological connection.

Father while master to his children. Due to Christ's bi-directional position, the master and slave find reconciliation through the mediator: humanity is reconciled to God by and through Jesus the worship leader.

This adds convincing validity for the relationship that Christ serves in being among the congregation and leading them in worship; he directs and leads from within. Because of the Son's death and resurrection, he uniquely becomes an heir with the suzerain, delivering the promised inheritance. Christ brokered the reconciliation that believers now can become children of the benefactor and brethren of the vassals. Therefore, the author illustrates that it is just as important to unify through the worship of Christ as it is to hold firm to the faith rather than relying on either the sacrificial system of the Jews or the common sacrifices for the purpose of business and social acceptance within the Greco-Roman culture. A worshiping congregation finds confidence in believing that the actions of Christ secured their ultimate reward. Because of this, Jesus is both the object worthy of worship as well as the source of worship itself.[11] For the modern church technical artist, they too are both suzerain and vassal; subject to both Christ and the mission of their church leadership and principal over the way the congregation will understand the worship message. In this way, there is a dual responsibility to ensure that the presented message clearly proclaims the message of Christ at all times while also being obedient to and aware of what that message is.

THE USE AND IMPORTANCE OF PSALM 22:22

Psalm 22 is the second most quoted psalm in the NT, only surpassed by Psalm 110. All references[12] focus on lament and suffering, except Hebrews, which emphasizes the glory and praise of deliverance continued from the Psalm 8 motif cited in 2:6–8. Michael Wechsler subtitles Psalm 22, "A Prophetic Perspective on the Crucifixion of the Messiah," which is an exact explanation of the circumstance when viewing the psalm from the standpoint of the exalted Christ rather than the words of David.[13] Psalm 22's two distinct sections—an individual lament (vv.1–21) and a song of thanksgiving (vv.22–31)—work in unity as a prayer song, following the common practice of the temple priests who begin with solemn prayer and shift to end with songs of

---

[11] There is an important distinction here, one which produces various possible objections. From a Jewish context, Jesus could not be viewed as one equal to God, however, a Greco-Roman worldview would require an equal status with a deity in order for the death to be made honorable. Greco-Roman emperors commonly took on the identity of a deity. Therefore, status of "God" would not be a concern. The author's goal is to merge the suffering honor of Christ with the God-Man identity in order to form an object-person worthy of worshiping, suffering, and enduring.

[12] Matt 27:35, 39, 43, 46; Mark 15:24, 29, 34; Luke 23:34, 35; John 19:24, 28; Rom 5:5.

[13] Wechsler, "Psalm 22," 778.

thanksgiving. The psalm opens with Jesus's cry of "my God, my God, why have you forsaken me?" (cf. Matt 27:46). Yet, on the words "you have heard me," it turns into an oracle of rescue. It becomes a confession of trust, moving from a cry for help to an outpouring of worship by both David, the original psalmist, and the author of Hebrews's congregation. Mark Heinemann calls the craftsmanship of Psalm 22 "extraordinary" due to its use of bold images and sweeping historical scope: David pleads for deliverance, acknowledging God had heard him, leading into praise and a call for the both Israelites and Gentiles (v.29) to seek the Lord as well.[14] Christ likely read this psalm many times, and it is, therefore, logical that at the cross he quoted this psalm in order to draw the messianic and prophetic connection. In turn, that allowed the church to adopt the entire psalm as its own. In this way, Psalm 22 the "theological center" of the book of Psalms.[15] The narrative begins with the search for the promised land yet ends in salvation of all people through the suffering Messiah. "Why have you forsaken me" becomes "I will praise you." For Christians, Psalm 22 is the message of the Messiah's humiliation and exaltation. At v.22, David's celebration is the embodiment of Christ's ultimate work on the cross, with Jesus becoming the singer of both lament and victory chant.

Psalm 22—the expression of David's dire circumstances—is, in fact, the most exact pre-Jesus description of the actual events that happened to Christ during his lifetime.[16] Though a direct correlation to the passion is not stated here, the epistle's audience would have likely made the connection. When the NT cites the OT, the broader context of the original meaning is also being invoked. Hebrews 2:12's citing of Psalm 22:22 would be the victory song of God's deliverance. Up to that point, the psalm is only abandonment, loss, and hurt. It becomes a rescue through pain and suffering by the only one who can truly atone for sin: God himself. The author expresses that because cross-bearing Jesus called out his forsakenness yet persevered, Christians can call out in praise of being found. The exalted Messiah will praise God to the people of Israel and in the congregation, calling on others to praise him (v.23). Closing the psalm, in vv.25–26 the Messiah calls the afflicted to seek the Lord and trust him for they shall "eat and be satisfied," having their "hearts live forever," offering salvation to the whole world. This viewpoint could not have been understood by David's congregation, yet would presumably be fully recognized by persecuted Christian Jews living in a gentile land.

---

[14] Heinemann, "Exposition of Psalm 22," 286–89.

[15] Kidd, *With One Voice*, 86.

[16] Even though some scholars contend that this psalm is merely David writing about his own suffering, Wechsler ("Psalm 22," 778–80) suggests that because of Christ's reference to it on the cross this is better thought of as a messianic psalm of the future savior fulfilled by Jesus rather than a description of David's actions or metaphorical feelings of struggle. David never had his hands and feet pierced (v.16), never had his clothes divided and lots cast for them (v.18), and never did his victory bring universal righteousness (vv.27–31).

By reading the psalm through a reflective worldview of modern Christianity, reading back the experiences of Christ, it may be possible to know more about Jesus—i.e., his emotions and prayers—than may be found elsewhere in the biblical text. The logic is that if it is possible to understand a majority of the text to be consistent with what is known to happen, then the conclusion can be drawn that it all did, even if there is not another direct parallel found within the NT. This includes not only the actions of Jesus but his emotional states of mind as well. This matches the messianic understanding of how God works through history: it is one thing for someone to write something about himself—David in this case—and have it coincidentally happen again to someone else, yet it is a whole other for nothing to happen to someone and it be a fully, uniquely, and completely fulfilled revelation in another. The details in Psalm 22 are so distinctive they could not truly have all happened to David in the way that they better match the exact details of Jesus's circumstances. Therefore, it could be said that the entire psalm must be known to be actual events of Christ, including his leading in praise and worship. I do think it is important to note that this view is not commonly shared by conservative writers of today; however, it has held considerable support from early Christian writers like Justin Martyr, Leo the Great, Eusebius of Caesarea, and Martin Luther.

In this way, the author's use of Psalm 22:22 in Hebrews 2:12 can be viewed to offer a parallel with worship in practice as Christ is now—in the present—revealing himself in the midst of the church proclaiming God as well as calling his people to respond in proclamation of God's name to the Father. Christ's role in worship is to lead people in worship of the Father through himself. The text then suggests that the suffering done by Christ is actually a thanksgiving because it opens the door to the Father as saved brethren-children. What was once suffering, is now triumph and solidarity between the Son and his children. It is the "surface argument" that exposes three main points: (1) part of Christ's role is to proclaim the name of God; (2) it tells the loci for which his name is to be proclaimed, namely within the congregation (ekklesia); and (3) the way to proclaim God's name is to "sing" praise.[17] In this way, it can be proposed that a proper response to God for the modern church—which includes the church of the Hebrews—is singing praise to and of God for deliverance from sin and into eternal rest and reconciliation. Worship becomes both present and participatory.

THE USE AND IMPORTANCE OF ISAIAH 8:17–18

Hebrews 2:13—Isaiah 8:17–18—is a support text for the ideas established

---

[17] Attridge, *Epistle to the Hebrews*, 104–5.

in 2:12. As opposed to Psalm 22, Isaiah was understood as messianic to the Jewish audience. Isaiah 8 is part of a more extensive section that focuses on whether the people will trust God or earthly powers, with these verses dealing specifically with the difference between a rebellious people turning from God and those who choose to walk in the way of the Lord. The exact date of original authorship is unknown, but the text was written from the viewpoint of an impending attack on Judah, presumably the 734 BCE Assyrian campaign into Israel. Which direction will they choose? Give into the world, or follow God? Rather than trusting God, the people turn to pagan superstition for hope. The author ridicules their actions and calls for trust in the Lord. As the people are increasingly being captured, Isaiah chooses to trust the Lord, even in the dire situation, holding on to the hope that peace will come. In the Isaiah 8:11–18 subsection, he is encouraged to remain separate from Judah and not follow after the ways of the world. There is a contrast between those who seek the Lord and do not fear him and those who do. Likewise, for some in the Hebrews congregation following Christ is a stumbling block, while other will fear the Lord and find refuge in him as a sanctuary in times of struggle.

In Hebrews 2:13 the author separates Isaiah 8:17–18[18] into two distinct statements, suggesting that two separate points are being made. The first point is to trust in God and his saving work through Jesus and not to focus on the frailty of the human condition. The second ties together the image of Isaiah's hope that Israel's "children" are signs with the children becoming one with Christ. Verse 13b comes from Isaiah 8:18 in the LXX where Isaiah speaks of his own children. Like with Bezalel and Oholiab, the reference to his earthly children as the testimony of trust in the Lord include names as a play on words and meanings: Isaiah means "Yahweh is salvation," and his son's names mean "a remnant will return" and "quick to the plunder, swift to spoil" respectively. This parallels the text in Hebrews because the people are not children of Christ due to their humanity but because of the work of God. The solidarity with Christ confirms the solidarity with the "children" of God, not because of being human but because they are the gift given to Christ by the Father. The conclusion is that Sonship is, therefore, not by birth—as would have most likely been thought through the seed of Abraham—but through the salvation provided by God through Christ. The Sonship through Christ is to be shared with his "brothers and sisters" who are proclaiming God's name through song alongside him as a gift from God. The human response is the singing and leading of praises to the Father. Whereas Christ's work is the reconciling gift from God to his children, being made brothers

---

[18] It is possible, as some scholars like Cockerill (*Epistle to the Hebrews*, 142–44) believe, that Heb 2:13a is from 2 Samuel 22:3, David's Song of Deliverance rather than Isaiah 8:17, because the image of David would have already been in their minds from 2:12. However, the direct, continuous connection between the two citations in one complete verse makes Isaiah a better fit, not separating the two halves.

and sisters of Christ is a mutual gift shared between Jesus and his people.

Isaiah is the author speaking, yet the words are placed in the mouth of Christ. Just as the Psalm 22 citation is at a textual turning point, so too is Isaiah 8:17–18, with each author finding sanctuary in the Lord. The waiting is fulfilled in Christ: a kinsman connection and not merely a discipleship model is established. It is not that followers of Christ are solely brothers, but they are also children of God. The use of Isaiah along with Psalm 22 demonstrates that by being in the midst of the congregation for the purpose of sanctifying them, Christ reveals himself as both the deliverer and one with the delivered. The use is not about the text's contents per se but about the relationship to what God has done through Christ. The brethren-children and Jesus are now of the same family; they suffer together. As his children, they are to be confident in their final glorification alongside the exalted Christ. The text then suggests that because of this salvation found through Christ, the proper response is praise and worship with the Savior.

PRINCIPAL THEMES

Five key themes are established in the text surrounding Hebrews 2:12–13, which all express Christ's unique identity. Together they support the reason and purpose Christ adopts the position as the Mediator of New Testament worship practice.

CHRIST AS FULLY HUMAN AND FULLY DIVINE

Moving Christ from man status to God status was an important step in establishing the Christology of the church. The two books that are the clearest on the deity of Christ—John and Hebrews—are also the ones that emphasize his humanity the most. Scholars formerly believed that a high Christology came from a late gentile or pagan influence, but the consensus today has it within an early Jewish setting, mainly due to how passages like Hebrews portray Christ in very distinctive Jewish terminology and within a Jewish context.[19] The pastor-author does not quote anything from Jesus's earthly ministry, only from the Septuagint; yet, viewed in light of Psalm 22 and Isaiah 8, many allusions to his earthly ministry are made by placing the OT texts as the words of Jesus.

The immediate text is one of the most important passages in the NT for the incarnation of Jesus as fully divine, eternal, and exalted. It is relational:

---

[19] Chester, "High Christology," 22–24.

Christ's position is lower than the angels, present with humanity, but with the world under his footstool, seated on the throne at the right hand of the Father. To the author, Jesus is brother, Son, child, God, sacrifice, and priest. God reconciled his brethren-children to himself as the sacrificial priest possessing the authority of God. The kingship of God creates the Lordship of Christ and kinship with humanity. Therefore, it can be suggested that the theocracy of the Jews becomes a Christocracy for the NT church.

Where chapter 1 establishes Christ in his unique role having worship of him purposed eternally, chapter 2 acknowledges Jesus's humanity because he is willing to identify with God's people from among them. Affirming the brothers and sisters corresponds to God's proclamation of Jesus's divine Sonship (1:5). By placing trust in God, Christ is accepting God's sovereignty as well as his ability to bring reconciled humanity with him to God's presence at his right hand (1:13). The summation of the OT quotations demonstrates Jesus's trust toward the Father, as well as an even greater meaning beyond the work of Christ's obedience to the Father: it is a promise to his brethren-children in the present. Because of Christ's trust, he can now become their worship leader in kind. He is both worship leader and the object of worship.

So then, to whom is worship sung? Worship is offered to the fully human and fully divine Jesus. God-Man is joined with his saved remnant as one family. His identity among them creates a song of encouragement because it demonstrates God's ability to relate to his people's hardships, disappointments, fears, and struggles. God-above becomes God-with. I propose that because the author transforms the OT verses ascribed to God and focuses them on God's announcement of Christ's deity to the Father, Jesus becomes equally worthy of worship. Using Psalm 22 and Isaiah 8 from the past in the present grants them eternality, moving from previous situational references to declarations of Christ as divinity-humanity. Being done so from within the church body suggests that the proper response to their relationship is worship of the Father through Christ alone. Christ's worship actions become the example of what his brethren-children are to imitate.

BRETHREN-CHILDREN OF WHOM CHRIST IS NOT ASHAMED

The chiastic nature of Hebrews 2:12–13 emphasizes the central point that "not only do I [Christ] commit myself in trust to God, but I present also the children whom he has given me."[20] The complementary use of "congregation" conveys Jesus as participatory in the worship experience among his brethren. In fact, every statement is offered from the perspective of a

---

[20] Ellingworth, *Epistle to the Hebrews*, 169.

personal affirmation: "I." Due to this, Jesus proclaims that he is performing the actions voluntarily: Christ is speaking to and for his brethren-children. In this way, v.13 states that believers see themselves as Jesus's object of declared trust. The verse adds a level of connection between the Father and humanity, which further supports Christ's role as mediator. Read in this way, Jesus becomes the "sole-tether" connecting God and his people.[21] Likewise, the term for "children" (*paidia*) is also commonly used in reference to younger people who are in close relation to the one speaking; therefore, being brothers, sisters, and children all work together to demonstrate the intimacy and the close relationship between Christ and his saved people.[22] Calling them children denotes their dependence on the Son. Just as Christ in his earthly ministry kept faith in the Father, so then should believers.

Verse 2:12 could have simply stated God's proclamation of his brothers and sisters if the goal were to show solidarity between sanctified and sanctifier, brethren and Christ, within Christ's human ministry. Yet, the homilist explicitly continues to state that Christ's work is done from within the assembly of the congregation, the same people he demonstrates his love for by dying on the cross. Jesus reconciles from among his brothers and sisters. In 2:14, the use of flesh is likewise important because it links the church as physical brethren of the same Father showing Jesus shared a common humanity with believers. It was his physical death that was able to atone for sin; for through it, Christ was able to sanctify—set apart, dedicate, and make holy. Though the verses are about Jesus's humanity, his connection is not anthropological but theological. Even though Christ descended from the promised lines of Abraham and David, the reference to "out of one" is ambiguous. It best fits the idea of being joined through God rather than his human Jewish lineage.

The NT church relates to Christ as his congregation-brethren. Because Christ is man-exalted, his dual status allows him to erase the shame of crucifixion. Through Jesus the brethren-children are made holy, which was formerly a status only held only by God himself. Accordingly, today, "the author would have the believer see himself or herself as the object of Jesus' declared trust,"[23] supporting Christ's claim that he is not ashamed to call them brothers. Jesus places trust in humanity to come to him and obey, voluntarily opening the way to the Father. For church technical artists, they should not be ashamed of their work and service because it is a work shared with the exalted Christ. Those who respond to the call will experience worship of and to the Father with and through Christ.

---

[21] Allen, *Hebrews*, 218.

[22] The use here is in the nominative, plural, neuter, denoting that these are male and female brothers-sisters-children This is similar to the use of *adelphos* used for brethren and accounts for both brothers and sisters.

[23] deSilva, "Responding to God's Word," 116.

TRUSTING IN THE FOUNDER AND PERFECTER OF THE FAITH

Beginning in v.10, Christ is presented as the perfect pioneer-founder-author of salvation. As brethren-children, believers are made perfect through the cross. The word *archēgos* is made up of two Greek words, *archē* (first) and *ago* (lead), giving the idea of "a leader who stands at the head of a group and who opens the way for others to follow."[24] The LXX uses the term to refer to those who lead the tribes of Israel in the wilderness and in battle (Num 10:4; 13:2–3; Judg 5:15; 9:44; 1 Chron 5:24; 8:28). In the immediate passage, 2:10 and 2:17 use *archēgos* in reference to Christ—just as it is used in Hebrews 12:2 and Acts 3:17. He leads to salvation in addition to leading the worship role alongside his brethren inside the church. The word "perfect" appears fourteen times in Hebrews. The author does not use it in the sense of morality or character but of completeness of God's plan and purpose. Christ, as the pioneer and perfecter in the midst of the congregation, provides congregational leadership to assist in enduring through trials toward the purpose of reconciliation.

The fact that Christ needed to taste death in order to be the perfect sacrifice necessitates that there was a purpose in his coming. Verse 11 expresses this reason: sanctification, "being made perfect." Sanctified believers are one with the sanctifier in brotherhood through the Father. On the cross, Christ's purpose is brought to perfection, ushering out the old covenant for the new. Hebrews places Jesus as the one who makes holiness a characteristic of those who are made holy, qualifying him to be the perfect mediator. The phrase "for it was fitting" ties Christ's action of becoming human with the ultimate purpose of final glorification for those who endure, which in turn serves as an encouragement to the author-pastor's suffering congregation (Heil, 2011, loc.986–93). In the Greek text there is a wordplay between "leading," "initiator," and "perfect." All three work together to highlight the bringing to completion. Early Christians most likely would have understood the messianic meaning of the call to endure. Thus, they too are the ones who would experience the perfect victory, able to find comfort in their worshipful praise.

Commenting on Hebrews 12:2—the parallel verse to the pioneer verse in 2:10—John Witvliet suggests that viewing the text in this light demands that it is impossible to have proper worship without Christ at the center: Jesus is the one through whom believers sing, commune, preach, and pray.[25] They do so because of both the believer's faith in God and Christ's faith in his brethren-children. Faith (*pistos*) is an important theme in Hebrews; in 2:17 Christ

---

[24] Allen, *Hebrews*, 214.
[25] Witvliet, "Joy of Christ-centered, Trinitarian Worship," no pagination.

is depicted as not only High Priest, but faithful High Priest. "Faithful" becomes his identity. It is who he is in delivering on the perfect promise. Jesus in his suffering still was faithful and trusted God. In his exaltation he proves his faith by remaining among his brethren-children, leading them to trust in worship of the Father in whom he places his faith.

This is an important aspect of the Christ-brethren relationship for modern church technical artists. In church tech circles, there is a common theme of daily suffering and enduring. For example, even the name of a popular website and conference, FILO, is a play on the identity of the technical artist; church techs are often the "first in" and "last out" of the sanctuary because of the time needed to set up, work out issues, hold rehearsal, run the service, shut down, and tear down. Technical artists claim to find themselves feeling underappreciated because their role includes jobs that the average congregation does not witness. In this way, the technical artist who trusts in the founder of their worship method would be honored as faithful, exactly for responding to the call to serve a ministry in which there is little acknowledgment or praise. Their worship practices are made perfect through their suffering and service. Thus, Christ finds their service "fitting" with his call for them.

MERCIFUL HIGH PRIEST

A major theme in Hebrews is the priesthood of Christ. Hebrews is the only NT book to refer to Jesus as the High Priest; it does so in the context of both the old and new covenants. Christ is High Priest after the order of Melchizedek (7:11), a priest who, like Jesus, was not from the Levitical line. Jesus's humanity among the people was only a temporary state; Christ as High Priest displays his eternal state. The argument for Christians is that because Jesus laid the foundation for salvation and sanctification through himself, it qualifies him to become the High Priest, representing his people before God. In all given duties, Christ embodies the high priest in every aspect except that, unlike other priests, in his humanity he was fully sinless, which provided for the perfect atonement that no prior earthly priest could offer because it was done so through offering himself as the pure and perfect sacrifice.[26] While the earthly high priest's sacrifices were to be repeated annually, Christ's obedient sacrifice of himself is permanent and eternal. As High Priest, he not only administers the sacrifice but is himself the sacrifice, showing himself as merciful (2:17; 4:15–16). Sharing in the people's human

---

[26] All sacrifices were to be clean and spotless which means that no human could be the sacrifice except if he or she is without sin in all aspects. It is this characteristic that qualifies Jesus for the position for Christians (Exod 12:5; Deut 15:21; Heb 9:14; 1 Pet 1:19).

experience, to believers, through Christ's death they are consecrated and made pure. The necessary response is worship of him and with him (Rom 12:1).

In Roman society, the emperor too offered sacrifices to the gods as the *Pontifex Maximus* (high priest), similar to the role of the Levitical high priest. In this same vein, the author presents Christ as the High Priest, viewed from both Greco-Roman and Jewish points of view. *Archēgos* (pioneer) was often used in classical Greek to refer to the hero of a city, by whom it is named and serves as its guardian, the one to whom and on behalf of whom worship and sacrifice was made.[27] In the Hellenistic world, because the fear of death was common, Christ's power over death—demonstrated through his exaltation—could have solidified Christ's function as eternal priest and hero. Thus, believers were not of noble stature as defined by Roman societal structures but through an atoned-for relationship with Jesus.

Besides administering the sacrifice, the Jewish priests would also lead the people in worship. Christ's identity as High Priest makes him the one supremely qualified to lead his people in worship. As merciful High Priest, he offers worship along with and on behalf of the congregation. Hebrews 7:25—a parallel passage to Hebrews 2:12–13—presents Christ as the enabler of worship so that followers can be reconciled with the Father through Christ alone as the intercessor for them. Thus, when the NT speaks of the union with Christ,[28] it includes following him to the Father through worship. In this role, Christ is tangibly in the midst of the congregation as eternal High Priest leading worship. Thus, if his sacrifice is eternal, so is his ability to worship the Father from his priestly position in front of and on behalf of his congregation eternally and present in the present. In this way, Christ is the earthly worship leader as well as the priest of praise in the eschatological hope still to come.

When Christ sings, he is not singing a solo but leading the choir.[29] The worship of believers becomes his worship. Combining the two roles of administering the sacrifice and being the sacrifice himself, contemporary worship can thus be viewed as necessarily sacrificial. Worship becomes worthy because of Christ's grace and mercy, not because of anything his believers have done but because he provided the way in his priestly role. The high priest of Israel in the Holy of Holies, represented the entire nation; Christ as High Priest represents all humanity who respond to God through him. Christ as sacrifice and worshiper in the present with believers, then, makes worship worthy because he calls his brethren to make their worship his worship. Hebrews 8:2 further verifies this by referring to Jesus as *leitourgos* ("leader of our

---

[27] McKnight and Church, "The Humiliation and Exaltation of Jesus," 70.

[28] 1 Cor 1:30; Eph 2:7, 10, 13; 2 Tim 1:7; 1 Pet 5:14.

[29] Man, *Proclamation and Praise*, 35.

worship"). Whereas Israel had priests who offered worship on behalf of the congregation, Christ himself became the mediator and leader of contemporary worship of God: he represents God to humanity and humanity to God in and through himself.

CHRIST THE APOSTLE

Jesus serves as both Apostle and High Priest (v.3:1). Apostles are charged to proclaim the good news, just as Christ proclaims of himself. What he asks for in an appropriate offering, he provides through himself. He draws believers near to him by being the proclaiming Apostle through priestly offering. He fulfills his own holy ordinances of worship, proclaiming the good news.

Jesus serves as God's messenger. As the apostolic mediator, he is able to represent humans before God. It is noteworthy that throughout the entire book of Hebrews, the only apostle the author mentions is Jesus Christ, even though the pastor-author himself most likely walked, served, and preached alongside the other biblical apostles during their earthly ministry. Possibly the author was even an apostle himself.[30]

Hebrews 2:12–13 illustrates the twin roles of apostle and high priest in microcosm. In the v.12a proclamation, Christ is revealing the Father, which was his actual earthly mission.[31] Accordingly, Hebrews 2:12a suggests that Christ's ministry of proclaiming the good news of the Father is continued today through the worship of the church.

THE SINGING SAVIOR

Hebrews 2:12–13 demonstrates Jesus's unique identity as Christ the "Singing Savior." The first Christians we know of who considered Jesus the Singing Savior was the church in Syrian Antioch (Acts 11:26). They produced their own psalm-like hymn book called the "Odes of Solomon," in which Ode 31:3–4 depicts Jesus singing, chanting, and lifting his voice to the "Most High" with them in their midst.[32] Clement was the first theologian to recognize Christ as a Singing Savior in Hebrews, writing that Jesus sings to those

---

[30] Issues of authorship are not the concern here, however, the evidence in the writing suggests that the author-pastor was tightly involved with, or at least influenced by, the Pauline circle. If the author is not Paul himself, he was most likely a disciple of Paul.

[31] Heb 10:5–9—a citation of Ps 40 that is also placed onto Christ's lips—additionally speaks to the ministry of Christ on earth. His earthly ministry is one in which Jesus never restrained his speaking or revealing of the Father to the people.

[32] Charlesworth, *Earliest Christian Hymnbook*, 92.

and for those he saved: Christ ties kinship to a worshiping congregation. He entered their human condition, sings with them, and leads them in worship of the Father through his saving work. Christ made the first part of Psalm 22 his own on the cross and takes on the second half in his resurrection. He expresses it musically, revealing his fellowship while praising the Father. Christ is present in leading praise to and of God; he is the leader and initiator of worship and praise.[33]

Christ's unique role brings to light four fundamental concepts: (1) the praising of God's name; (2) singing to the Father through Jesus; (3) his position in the midst of the congregation; and (4) revelation and response. I propose that all four positions work together to formulate how Christ's position plays out in contemporary worship.

PRAISING HIS NAME

During his earthly ministry, Christ proclaimed God's name in the synagogues.[34] In the book of Hebrews, he proclaims it to his own reclaimed and reconciled people. The name of God is the Father with all his attributes, nature, character, conduct, and saving action, proclaimed as the chief liturgist. Therefore, God's name is more than just who he is but includes what he has done for his brethren-children. Both 2:12 and 2:13 cite an OT verse which transitions from lament to praise and thanksgiving. It is the vindication after suffering. In this sense, being made brethren is the joy that believers feel because of the suffering and saving work of Christ. Here Christ takes on the role of the singing priest, leading his people in praise. Thus, Hebrews demonstrates that Christ and his people praise God in union with one another; there is no other purpose than praising the name of God.

Worship takes another step; worship is recognition of his name—his identity—in its entirety. Worship through Christ makes a praise offering perfect. To the early church, Christ was not the object of worship as contemporary believers treat him but instead was the subject of praise and the content from which worship was constructed. Worship would not be done at Christ, but about Christ. This is significant because, while modern Christians generally have no problem worshiping Christ, if the early church did not view Christ in the same light in terms of worship practice, how they would have understood worship to work within their Greco-Roman and Jewish circles

---

[33] Some theologians have tied this to an eschatology, noting there are two simultaneous worship activities: earthly worship that passes away and the heavenly, exalted worship that believers are guaranteed through the saving work of Christ. Christ's earthly worship to the Father among his people can be viewed parallel to the heavenly worship found in the book of Revelation, where all of God's people worship to the exalted Christ at the right hand of the Father.

[34] Cf. Matt 4:23; 13:54; Mark 1:21, 39; 3:1; 6:1–2; Luke 4:15–16, 44; 6:6; 19:47; John 18:20

would inform how they would apply worship practice to their newly formed and continually reforming Christian liturgies. They would have musically shared their adoration and praise of him through borrowed current, common song styles and worship structures. When the song is then performed about Christ-exalted, it becomes a victory chant over sin through the saving work of the Singing Savior.

In 2:12–13 Christ speaks to the Father on behalf of and for the benefit of his brethren-children, taking his place among the congregation. He does so through an active, present, and continual promise of worship. The God who is worthy of worship himself joins his people in worship. The word for "I will praise," *hymneso*, offers the idea of praise in the form of a hymn rather than simply spoken words. In this light, John Paul Heil suggests this serves as a form of speech act—a lyrical speech act of sorts—in which communal worship accomplishes God's promises for the benefit of the congregation.[35] The people are beneficiaries of Christ's work because from within the body arises both the sacrifice and Savior. Thus, too, from within the congregation comes the true Worship Leader and Mediator of that worship. In v.12b the praise transitions to specifically note that Christ will sing the Father's praise from within the assembled congregation. Ron Man deduces the reasons this should be understood as Christ himself singing in the church: (1) the use of *ekklesia* suggests that the author has the present church in view; (2) "brethren" and "assembly" are paired with one another in the same verse that has the point of Christ identifying with his people, consistent with the previous citation in v.11 of those who are sanctified; and (3) it is consistent with the original meaning of the thanksgiving section of Psalm 22.[36] This classification of Christ leading from within the congregation is significant, because it shows a distinct change from former Jewish practices wherein YHWH was unreachable to the average Jew. For NT worship, Christ is not only reachable but present with the believer in worship. If viewed in this light, the conclusion could then be drawn that for the modern technical artist, their construction of worship from within the church too is led by and with Christ from among.

SINGING TO THE FATHER THROUGH JESUS

In John 14:6 Jesus pronounces that he alone is the way to the Father and "no one comes to the Father except through me." Hebrews 2:12–13 presents a tangible way for God's brethren-children to reach the Father in worship: through Christ. Christ's humanity reconciles the congregation to the Father, and because of that, they are brought together in worship of the Father.

---

[35] Heil, *Worship in the Letter to the Hebrews*, 1036–38.

[36] Man, *Proclamation and Praise*, 29.

When believers sing to Jesus, praise is transferred through Christ to the Father. While wholly acceptable worship in a fallen state is not possible because no fully blemish-free sacrifice could be offered, the passage suggests that purification through Christ sanctifies worship practice. Jesus's mediatory role in performing the justification of humanity makes congregational singing the praise of victory that could only come through his sacrificial death.

John Witvliet summarizes this intercessory purpose: commonly when a person focuses on one thing, they take away from another. Worship through Christ necessitates focus on the Father, relies on the indwelling of the Holy Spirit, and is the gift of Christ from the Father.[37] It is holistic. James Dunn adds that even though the way to the Father in worship is through Christ, believers must be careful not to fall into "Jesus-olatry."[38] They must not throw out the Father in place of Jesus. Worshipers must remember that Christ is the vehicle, but the Father is the destination. The Spirit is the motivator. There is a common tendency today to assume worship of Jesus alone, yet the NT refers to Jesus as the icon (*eikōn*) of the invisible God. Thus, believers are reconciled to the Father through Jesus, and the worship due him ought not to stop at Christ. Jesus is mediator. As mediator, Jesus is the bridge between Creator and creation, between a fallen status and being his redeemed children.

IN THE MIDST OF THE CONGREGATION

Hebrews 2:12b places Jesus "in the midst of the church congregation." It signifies a dwelling place among the people of God, which in the NT world is the church, God's people. There is a sense of familial relationship that is personal, with members of the body coming together for one another. There is a particular physical dimension present in the use of "congregation." They are a people who choose to be present while at the same time are chosen ones, saved by God, who likewise demonstrate their praise of God's saving work with and through Christ. Psalm 22:22b's reference to the church means that it is a realized eschatology, not future or metaphorical.[39] This validates Christ's present role among his people because it suggests that what God promises in the future, he achieved in the past and continues in the present. That it happens "in the midst of the church congregation" reinforces that this is not only proclaimed in heaven (12:23) but on earth as well.

---

[37] Witvliet, "Joy of Christ-centered, Trinitarian Worship," no pagination.

[38] Dunn, *Did the First Christians Worship Jesus?*, 147.

[39] Ellingworth (*Epistle to the Hebrews*, 167–68) connects the reference to the *ekklēsia* from 12:23, citing that what is in view here is the heavenly Jerusalem as a worship that is participatory: what believers do in praise and response on earth is what they will do and be in their future eternal dwelling.

Early church writers, like Ignatius of Antioch, suggested that singing as one body of Christ was important to early Christians because the act itself brought about unity and identity: "Jesus Christ is sung" when the church is in accord.[40] Likewise, the modern church is in accord because Christ is among the assembly, leading and participating, as echoed in Christ's call for unity in John 17:20–23. The activity of singing is the "sounding image of the unified church"[41] which assists the brethren-children to grow in their understanding of both Christ's identity as well as the Father's while at the same time becoming united with others in the congregation. The song Jesus gave the people was a song of strength, encouragement, and endurance to push forward, since one of the issues facing the author's congregation was drifting away from the faith. Christ's singing "in the midst" supplies the reason not to fall away: Christ is building up the body for his ultimate purpose alongside them. He is doing the work through the congregation—for which technical artisans possess a special role as translators of that message.

Worship is not perfected due to the voices or skilled playing of the congregation but because Christ in their midst perfects it. The word *ekklēsia* was a common term relating to Jewish gatherings and civic gatherings in the Greco-Roman society. Christians took it to identify the church. In the other Hebrews verse referring to the church, the author uses *episynogogē* (10:25), the word for the physical church. "Assembly" is also mentioned in 12:22–23 with *ekklēsia*, our common use for church. Thus, the church is not a place but a people. The phrase, "in the midst of the assembly," is tied to the way Christ and early Christians would speak in the temple and synagogues. Even as a persecuted people, they would have everyday practices found in the local synagogues, with some possibly still having business and familial connections to Jewish practices. Jesus in their midst, thus, usurps Jewish practices for his church. This idea, though rarely mentioned in modern scholarship, was the common understanding of the early church up through the contemporary "traditional" hymn-writers. Jeremy Begbie footnotes Calvin and Wesley, citing that Calvin referred to Christ as the chief conductor of hymns, and that Wesley saw Christ as the conductor of the present-day choir, just as he did with apostles at the Last Supper.[42] Calvin understood Christ's presence among his people as a call to teach the gospel. Being among God's people creates a responsibility to encourage God's people and share the good news, openly sharing their gratitude, stimulating one another, and fervently singing praises to God as he acts as the chief composer of the church's hymns. In this way, as the people of God seek Christ and respond through song, they share Christ's identity among the congregation for the good of the

---

[40] Guthrie, "Wisdom of Song," 384.
[41] Ibid., 385.
[42] Begbie, *Resounding Truth*, 318n38.

congregation. It is through God's people where Jesus can be found.

Since the church is its people, not its building, for church technical artists Christ would likewise dwell in the midst of where God's people gather to seek and share their praise of God—in this case the sound booth, post-production room, control room, camera stations, greenroom, backstage, and catwalk, among others. The text suggests that being in their midst demands that where God's people gather to share the things of God, he is present "composing" their techniques and methodologies. Therefore, presenting Christ through sound, light, visuals, ambiance, and atmosphere are all external expressions of the same internal motivation as outbursts of song and vocal praise: all of which mediate the salvation message of Christ from the Father to the congregation.

REVELATION AND RESPONSE

Worship is about God revealing himself and his people's response to it. It is revelation and response. It is bi-directional, active, and sacrificial. Christ sacrificed himself to be the saving revelation of God in order for believers to respond to the Father through him. In contrast to Psalm 22 which moves from the aloneness of the singer to the praise among the people, Christ's earthly ministry was among the people, while the atoning sacrifice was solitary. The victory is public. Yet, like the psalm, NT worship begins as a solo but ends as a chorale; God alone reveals and his people in unity respond. It is a synthetic and opposite proclamation. In v.12a it is God to humanity—announcing God's praiseworthiness to the people—and v.12b is humanity to God—the appropriate praise lifted to God in response. The practical application of the response bears heavily on those who have the task of presenting the Bible to the congregation through the technical arts because they are Christ's representatives mediating the response to God's saving work. Mirroring the method found in Hebrews 2:12–13, for the church technical artist, it is done through both participating in worship through receiving and leading others through artistic translation. Thus, the text suggests that acceptable worship would, therefore, incorporate both revelation and response.

The concept of revelation and response mirrors early church practices, which centered around singing, creating, teaching, and communion—all acts of reverent worship. Additionally, music in worship would serve as either presentational or participatory. Since ancient Hebrew worship included the use of singing hymns, worship practices that included singing would be a natural characteristic for the NT church to carry forward. As a tangible example, early Christian practices included call-and-response songs where one person would sing a line and the congregation either repeats or affirms in amen.

In the revelation, Christ pronounces his dual role. Jesus sings a hymn to the God who answers his cry. Where the Gospels present Christ's earthly ministry as the mediator of the Father to the people,[43] in Hebrews Christ performs the task from his exalted state rather than his former earthly one.[44] When contemporary Christians sing and teach they must be aware they represent Christ and his ministry of revealing the Father. When Christ leads and sings, it is not in isolation; it is in the midst as High Priest, offering an outpouring of God's saving grace.

In the response, Christ not only mediates; he also participates. He demonstrates the proper response. Christ as God—incarnate and exalted—deserves and receives worship, but he is also a worshiper himself. Because sacrifice is offered through Christ, praises are likewise delivered through him. When Christ praises the Father, he leads the way for worship of the Father in return. The people's worship is Christ's worship. His perfect offering takes imperfect worship and makes it perfect. Jesus is the agent, not an observer. Because God takes the initiative to reveal himself, believers are obligated to follow up with an acceptable response, demonstrated through both worship and obedience. In this way, God ensures his relationship and faithfulness to his chosen people. In worship, the response is bodily in action, performed through hearts and voices. Thus physically, singing is a natural act in response to the gift of salvation. It is a heart pouring out thanksgiving. For the technical artist the response would show itself through the physical act of directing sound and light to enhance the corporate praise offering to God through Christ.

WARNING: THE INSUFFICIENCY OF HUMAN MEDIATION

This chapter asserts that church technical artists serve as God's human mediators between the stage, congregation, and himself. However, it cannot be overlooked that the author of Hebrews directly argues against the adequacy of human mediation. Christ alone is presented as the sole suitable mediator, while the worship offerings of Aaron, Moses, and the high priest are cited as insufficient (Heb 8:3–6; 9:11–16; 10:1–4). The distinction is that the author is not arguing against their ability to serve as mediators but rather their ability to create permanent reconciliation between God and humanity. Christ's eternality makes his mediation sufficient. Earthly mediators could only atone for sin either annually or upon sacrificial atonement, while Jesus could atone once and for all. Aaron having to atone for his own sins

---

[43] Cf. Matt 11:27b; John 1:18; 3:34a.

[44] I suggest that Christ hinted at this in John 17:26 where he states that he made God's name known and will continue to do so.

demonstrates his insufficiency and the ineffectiveness of the old order. John Peter Lange explains the inadequacy in terms of worship:

> Under the authority of the Mosaic law and worship, there was indeed a calling to the eternal inheritance of the children of God; but the promised inheritance could not be received, because the law was able only to sharpen the consciousness of guilt, and with this the sense of deserved punishment and death, while the ritual could, in its turn, produce only, as a Levitical purification, a typical redemption, a merely symbolical approach to God. It was only through the truly expiatory death of the God-man . . . that a change was wrought in the entire relation of humanity to God, and a real taking away of man's guilty condition and relations became possible.[45]

In this way, it could be argued that—contrary to the claim that any mediation not performed in and through Christ would fall short—any mediation at all not performed by Christ himself would be insufficient, or at minimum a misdirection away from the true mediator. Even while God voluntarily chose to be among his people, it took Christ to perfect the worship offering.

It could be argued that the need for a "true tent" found only in the person of Christ as *leitourgos* suggests that no physical space is sufficient and any created location for worship is solely a representative "feeling" of connection, and not a genuine personal connection with God.[46] The author describes the earthly dwelling as the "bad example" and "copy" of the heavenly dwelling that is corrupted by human interpretation and sin.[47] Attempts to bring God into the sanctuary rather than use the space to point to him would be to use the modern elements of worship in the same way as Old Testament priestly attempts. "The earthly sanctuary provided no access to God 'but Christ' . . . [transitioned] the theme of sanctuary to sacrifice . . . show[ing] that no access to God was possible through its ministrations, . . . [but only with] Christ's sacrifice" of himself.[48] Just as the high priest was presented as holy to the Lord, anointed, and adorned with a golden plate and skilled needlework (Lev 8:7–9), his actions could not create a personal relationship with God and his people. The "show" fell short. In this way, no technical production can be a substitute for the reconciliation with God that only God himself can create. Technical artists who view the product as the end—rather than a means to the end—fail to mediate, becoming a representative high priest in the order of Aaron rather than as a means to connect God to his people through Christ.

Technical artists speak of "tricks of the trade" that make the audience "feel" a connection to God and the Holy Spirit. Examples include progressively dimming the lights as the worship set progresses in order to create a

---

[45] Lange, *Hebrews*, 159.

[46] Likewise, Exodus's mention of two tents could suggest the incapability of one tent to be sufficient.

[47] Michaels, "Commentary on Hebrews," 389.

[48] Cockerill, *Epistle to the Hebrews*, 387.

more intimate and reverent mood, playing keyboard pads underneath the prayer to generate a spiritual connection and make the prayer "sound" more transcendent, or repetitively displaying only a few words of song lyrics at a time laid over live video of the worship pastor singing in response in order to make people have to say every word rather than follow along in Bibles or hymnals. Ron Rienstra argues that such practices create "celebrity" out of the production because the effect of practices like these is more than simply techniques used to help the congregation worship; they convey their own meanings often contrary to the intended object of worship.[49] Indeed, the text appears to warn against such practices. "Making someone worship would be far more invasive than merely instilling a belief."[50] Forcing someone to worship diminishes the power of the object worthy of worship, making the worship experience unfitting. "One might pretend to worship God by following certain religious rituals, but that does not mean that one actually worships God."[51] These production tricks risk creating mechanical participants assuming a feeling of connection with God rather than learned worshipers.

According to the author, another reason human mediation is insufficient is because fallen people have the propensity to serve their own fleshly desires rather than seeking after God and trusting in him. This is the essence of the fifth warning passage found in Hebrews 12:14–29. The congregation was facing apostacy, desiring to return to their former rituals which would have been physical and emotive rather than spiritual and founded in truth. In this way, according there is a hermeneutical function wherein the theological truths being presented are stripped from their illocutive context and exchanged for the perceived emotive value of the artist. Thus, if modern technical artists are indeed the least theologically trained participants in the creation of the worship service there exists the prospect for misguided interpretation of the theological principles being presented when the apostacy is mirroring practices of the secular world of entertainment. It can be asked, then, how can someone who is not learned in a theological principle "add" to the multisensory experience in a way that highlights the aspects important to the preacher or worship pastor that he or she may not altogether understand? For example, Todd Farley questions how the modern church that appears dry and uninspired compete with MTV antics and the Hollywood talent pool that their congregations are pulled toward without becoming like them?[52] His answer is *actio divina*: God's self-performance through the art and artist.

Therefore, the production becomes acceptable when the artist is a changed person in Christ, allowing truths of the faith guide the artistic decisions made. Ron Man asserts that technical artists ought to:

---

[49] Rienstra, "Audio Technology in Worship," 27.
[50] Smuts, "Power to Make Others Worship," 231.
[51] Brown and Nagasawa, "I Can't Make You Worship Me," 142.
[52] Farley, "Theatre in Liturgy as Actio Divina," 33–34, 38.

strive for excellence in our worship, but not see technical expertise or artistic merit as ends in themselves, or as a means to gain God's favor or acceptance, . . . [because] ultimately our worship is pleasing to God only because we come through Christ.[53]

In this way, the act of generating a worship experience can appear contrary to the Hebrews text when acknowledging that all human attempts are temporary, require repeating, may reflect the desires of the human mediator over God's, and may create impressions of worship experiences over true connections to Christ. For these reasons the author cites human mediation as insufficient. Thus, mediation through Christ can only be done by Christ, and therefore the role of the technical artist would be to point to Christ, allowing his work to spur the congregation, rather than attempting to become a replacement for him.

CONCLUSION

Contemporary churches—as Christ's representatives in the world—mediate the gospel to their respective congregations through worship practices and the technical productions. This view holds that just as Christ is presented as the Perfecter of the faith who mediates worship between the Father and his people—Christ's "brethren-children"—technical artists fulfill a real-life role as mediators within modern church services, facilitating the message being delivered from the stage and pulpit to the audience via a skilled sound, lighting, and visual practice. In this way, the technical artist fulfills Jesus's model for worship, both as worshiper and worship leader. Jesus conducts from his exalted state, yet he is present from the midst of his church congregation; church technical artists conduct through interceding between the spoken and musical word delivered from the stage passed through their equipment from among and to the congregation. In this way, the congregation is lead in worship of the Father mediated through Christ's earthly representatives, church technical artists.

---

[53] Man, "Biblical Principles of Worship," no pagination.

❧ 5 ❧

COLOSSIANS 3:16: THE USE OF MULTIMEDIUM "PSALMS, HYMNS, AND SPIRITUAL SONGS" TO TEACH AND ADMONISH ONE ANOTHER

3:16   Let the word
            and message of Christ
        dwell richly among you,
           in all wisdom
        teaching
        and admonishing
            one another
      through psalms,
         hymns,
          and spiritual songs,
        singing to God
           with thankfulness
              in your hearts.

INTRODUCTION

Colossians 3:16—along with its parallel verse Ephesians 5:19—form the final pillar in this biblically informed view of technical artistry. Buried within the seemingly unrelated put-on put-off narrative is the first written example of early Christian worship practice and Christian hymnody. It demonstrates that the content and character of the early church was that of a singing church.

Like the previous two chapters, the musical and artistic aspects of the verse are often overlooked in theological journals and writings,[1] with the scholarly discussion focusing on the surrounding imperative actions for Christian living, which on the contrary has a had a vast array of discussion. The epistle summarizes what a Christian life is to resemble. It is a putting-off past behaviors and putting-on a new humanity that is realized through a personal relationship with Christ due to his prior act of grace. In this section, however, the commands are for the church body, a corporate act of worship and obedience. The respective commands to "be filled with the Spirit" and "let the message of Christ dwell richly" are followed by speaking, teaching, admonishing one another, singing to God/the Lord, giving thanks to God, and submitting to one another. All of which are done in the name of the Lord.

The principal theme of the letter is to establish the supreme authority of Christ, writing urgently in the wake of an attractive yet false teaching that is penetrating the church's congregation, known to modern scholars as "the Colossian heresy." Paul focuses on one particular teaching and argument in this letter, and even though he does not spell out exactly what that is, it seems to be clear to the audience. From that, we can draw conclusions as to the nature of what Paul's concerns are as well as how Christians are to behave regardless of the exact struggle they are facing. Paul desires to turn the congregation—whom he has never met—away from the false teaching and to encourage them to live a Christ-focused life. Paul offers the practical applications necessary to fend off this issue at hand.[2] One of the methods is worship. In Colossians 3:16 Paul uses music to describe a contrast between those living an old and new life. The people are to use psalms, hymns, and spiritual songs to admonish each other, put to death earthly passions, and instead be filled with Christ's qualities.

The verse demonstrates that when Christians gathered, they did not only pray and break bread but voiced praise to God through song. Barry Joslin

---

[1] Sasser (*Worship*, 117–19) comments that he found no direct commentary throughout history aimed specifically at the verses pertaining to psalms, hymns, and spiritual songs. While Pliny, the Ante-Nicene fathers, St. Augustine, and St. Paulinus wrote on the importance of singing in the church, this itself was not explicitly exegeted. In fact, the only resource discovered throughout the research process that was fully dedicated to a comprehensive exegesis and commentary of this phrase was James Janzen's, *Psalms, Hymns and Spiritual Songs: A Road to Unity and Spiritual Maturity* (2015).

[2] The authorship is often disputed, with valid arguments on both sides. Neither affects the outcome of this theory. Therefore, Pauline authorship will be presumed, with a later writing from Roman imprisonment. To question Pauline authorship is recent, and no early Christian (for which we have any evidence) doubted Pauline authorship. It was not doubted until the nineteenth century. Non-Pauline authorship has not found wide acceptance because the alternatives fall short of demonstrating strong arguments to being anyone other than Paul as the author. Though there are distinct possibilities that an associate of Paul was the author, there is more evidence to show it as a growing work of a matured Paul, seeking to instruct the church. This is the position held here. Paul, having completed a life of ministry, and now imprisoned in Rome, could easily have adapted his teaching to relational issues of the developing churches. Whereas his early ministry focused on building on Christ as the Jewish Messiah, his later ministry focuses on tangible applications to young churches throughout the Diaspora.

notes, "The command to sing is the most frequently repeated command in all of Scripture."[3] Even though citations of music in the NT are scarce, singing would have been part of the culture of the early church. Musical liturgy was active in both private and corporate worship. Douglas Moo suggests the passage demonstrates three core aspects of early Christian worship:

> First, the message about Christ . . . was central to the experience of worship. Second, various forms of music were integral to the experience. And, third, teaching and admonishing, while undoubtedly often the responsibility of particular gifted individuals within the congregation . . . were also engaged in by every member of the congregation.[4]

Paul tells the congregation how they are to accomplish this: he connects the singing of psalms, hymns, and spiritual songs to the rich indwelling word of Christ, the indwelling of the Spirit, the teaching and admonishing of one another, being wise, and the giving of thanks to God in their hearts. In this context, music has the distinct purpose of edifying, encouraging, and exhorting. Thus, the giving thanks to God, even musically, is not enough if the body of Christ is not built up and supported by one another.

> Christian congregations are to sing. They are to sing a variety of songs that teach in a meaningful and clear manner, theologically accurate doctrine. Such singing is to be understandable so to be directed both to God through Christ . . . and to 'one another.' Such worship calls for each congregant's involvement, not his or her participation as an observer.[5]

Paul appears to suggest that a church that is living out its true intended purpose for Christ will be a singing and artistically devoted church. The implications presented then can have a direct correlation to the artistry performed by modern church technical artists and that they too fulfill the role of teaching and admonishing the church body through administering the performance of their craft.

TEXTUAL CONFUSION

There is very little agreement among scholars on how Colossians 3:16 is to be translated, with no two popular translations being the same. The problem with the verse is the question of syntax—how the two halves, starting with "let the word" and "singing," respectively, relate to one another—which in turn affects how psalms, hymns, and spiritual songs is perceived. There are three main variants among translators: (1) each section serving as an

---

[3] Joslin, "Raising the Worship Standard," 50.
[4] Moo, *Letters to the Colossians and Philemon*, 290.
[5] Crabtree, "Colossians 3:16," v.

imperative; (2) each being an attendant circumstance; or (3) as a means by which something is to happen.[6] The difference is important because a differing translation drastically changes the meaning and referent of "word of Christ"[7] and thus how Paul's intended purpose for the use of music and worship is understood. Yet, all translations—without exception—demonstrate that the Colossians were a worshiping community. Considering both a literal translation and the use in context, however, does affect how worship is understood in practice: whether it is how the congregation would increase in their knowledge of God or whether worship is merely a liturgical act.

Common bias often separates psalms, hymns, and spiritual songs from the actions of teaching and admonishing as if they are distinct disciplines.[8] Even though the phrase "psalms, hymns, and spiritual songs" seems as if it should connect to the word "singing" that follows, it better fits with the previous "teaching and admonishing." The text naturally suggests that psalms, hymns, and spiritual songs are means by which one can teach and admonish, and in turn have the "word of Christ dwell richly" among the people. Those who disagree with this understanding, and that prefer each section to be an imperatival command of the word dwelling, appear to forget that the verse here is communal and not individual.[9] If it were individual, indeed, "dwell richly" could not be something a believer could make happen himself or herself; however, as an effect of an action, the message of Christ can further dwell among the congregation. Likewise, to place the latter "singing" with psalms, hymns, and spiritual songs takes doing some creative rearranging of the text. In fact, to make singing to God imperative is to add an "and" that

---

[6] This can be either modal or instrumental. Modal adds a sense of emotional color to the action. While some like Detwiler ("Church Music and Colossians," 347–48) suggests instrumental—simply the way in which something is to be done—is the usage here, the "actions" described are all emotional in nature: singing, teaching, admonishing. The best understanding is therefore modal. Joslin ("Raising the Worship Standard," 54) would concur, stating it is best to view them as modal, or "more clearly, adverbial participles of means describing how the action of the imperative finite verb is carried out."

[7] By "word of Christ" Paul meant either the spoken word (subjective genitive) or the teaching of Christ (an objective genitive), or possibly both. The overall idea of the letter does favor the objective genitive. Detwiler ("Church Music and Colossians," 351) suggests that if you cannot tell which one it is, maybe it is both. This fits both the character and action of the word of Christ. The early church would have likely been taught direct quotations of Christ as well as possessed memories of the message Christ delivered, even if delivered through the apostles or recent disciples. In the context of Paul's writing, he exhorts the congregation to recognize the superiority of Christ. Therefore, the message of his identity plays a central role in the epistle.

[8] Bruce (*Epistles to the Colossians*, 152) demonstrates this bias out of personal belief rather than accepting Paul may have intended an uncommon—to the 20th century—meaning: "It makes better sense if the phrase 'in all wisdom' is attached to 'teach and instruct' (not to 'dwell richly') and the words 'in psalms, hymns, and spiritual songs' modify the verb 'singing' (and not 'teach and instruct')." In this quote, Bruce is suggesting a translation based on "better sense" to him rather than accepting a possible alternative meaning that better fits the context when viewed in light of a worshiping congregation.

[9] The participles are introduced by a present tense prepositional phrase that denotes a contemporaneous action. Detwiler ("Church Music and Colossians," 355) cites D. A. Carson: "Computer studies of the Greek New Testament have shown that although a participle dependent on an imperative normally gains imperatival force when it precedes the imperative, its chief force is not normally imperatival when it follows the imperative." Thus, it is likely not three imperatives.

is not in the text. Because the "and" is not there, the participles should not be equal but a further description of means by which it happens. Therefore, with the second half of the verse, "singing" should be subordinate to "teaching and admonishing." Singing with grace and thankfulness to God is the way psalms, hymns, and spiritual songs ought to be performed in order to teach and admonish. A means by which understanding provides a balance between the two clauses that is natural to the text without modifying, adding, or rearranging it. It then matches the exact structure found in the parallel verse of Ephesians 5:19. In both texts, "Paul wants the community to teach and admonish each other by means of various kinds of songs, and he wants them to do this singing to God with hearts full of gratitude."[10] A closer examination of the verse finds that the text distinctly states that teaching and admonishing are through psalms, hymns, and spiritual songs and not through singing. Not that psalms, hymns, and spiritual songs cannot be sung—as indeed they most usually are. However, the text signifies that singing is not the only means; other mediums are likewise possible. All artistic endeavors are allowed here, including today's modern art form, technical artistry.

Scholars commonly break down Colossians 3:16 into participle sections, which leads to creating separate meaning for each individual selection rather than viewing the phrase in its entirety. This causes the focus on "let the word of Christ dwell in you richly" to have a meaning that the whole verse is not necessarily stating. The entire context proposes that there is both a musical and artistic aspect. Regardless of punctuation or sentence structure, the verse verifies that singing and liturgical worship are methods for mutual edification as well as vehicles for praise to God.[11] In this way, it can be suggested that believers are continually filled with the Spirit by coming together in song and worship.[12] Likewise, the verse possesses a present tense participle with a present tense main verb: the action is present and ongoing. Worship is to be a continual means for teaching, admonishing, and living the message of Christ among the congregation.

EXAMINATION OF THE IMMEDIATE CONTEXT

Colossians 3:16 is a letter to gentile Christians that belongs to a "two-

---

[10] Moo, *Letters to the Colossians and Philemon*, 288.

[11] For the sake of this chapter, I default to citing "singing" because it is in the text. However, it should be understood per my conclusion from the previous paragraph that all artistic forms of worship are in view, which includes technical artistry.

[12] There are obvious theological debates regarding how the Spirit works in one's life, whether the filling of the Spirit is a one-time event or a continual action. With the context and literary usage in this case, Paul is not presenting a view from the point of baptism of the Spirit upon salvation but rather a continual filling necessary for leading the Christian life. Not that Paul would disagree with the filling of a believer upon conversion, this is not the context being stated here.

ways" form of literary instruction, where both positive and negative actions are compared and contrasted in order to show the proper way of living. The verse is set within the minor "put-on put-off" section of 3:5–17. It presents warning and advice that focuses on lifestyle implications of the shared experience that believers are to be full of thanks and gratitude toward God for what he has done, causing them to live in peace and forgiveness of one another. The surrounding text contains fifteen commands, all seemingly with equal weight, with four being "stop" practices and eleven being "do" practices. While scholars often see the inclusion of psalms, hymns, and spiritual songs as simply a miscellaneous "stray mark" within the pericope with little meaning to the surrounding verses, a complete reading of the epistle shows that it is no stray mark.[13] The text suggests after one "puts off" the bad practices, he or she is to "put on" love of one another. One method of doing so is through the performance of psalms, hymns, and spiritual songs. Paul follows this section with tangible methods of submission in personal relationships: wife-husband, children-parents, and slaves-masters, suggesting that what is gained from the corporate worship is to be practiced in private affairs.

The theme of the supremacy of Christ is woven throughout the letter, with the central point being the unity of all things. It is explicitly stated in 3:16 in regards to Christ's message needing to dwell among the congregation, while being done so in the name of the Lord (v.18)—which would be at odds with the gnostic concepts being presented by the Colossian heresy. The epistle is a theological treatise hidden in letter form. In Colossians and Ephesians Paul focusses on the church body with Christ as the head where the other Pauline letters emphasize the Spirit. Colossians 3:16 and Ephesians 5:19 together link an active living out of the faith with the Spirit and message of Christ in a way that joins the church to its members while revealing how a higher Christology works within the church body. The pericope of 3:14–17 places the focus on a new goal: love. This follows well Paul's citing of psalms, hymns, and spiritual songs within the context of living holy lives. In this way, worship inspired by the Spirit involves giving thanks, singing psalms, and congregational discernment.

As a literary textual examination, Colossians is a work of art that interweaves various literary forms and genres, with 3:16 directly relating itself to the style of the letter itself. It states that one should employ psalms, hymns, and spiritual songs in order to teach and admonish, and Paul does just that with the various literary styles and references found throughout the entirety of the letter. The epistle includes thanksgiving and prayer reports, hymn, vice and virtue lists, household codes, and general exhortation, with the Christ hymn of 1:15–20 serving as the Christological high point. Thus, the singing of psalms, hymns, and spiritual songs to teach and admonish is not a far-

---

13 Guthrie, "Wisdom of Song," 386.

fetched idea, considering Paul himself uses a hymn to present his hortatory speech.

THE FIRST-CENTURY CHURCH

Colossians 3:16 demonstrates that early Christians believed an essential element of liturgical practice was not only breaking bread, prayer, and the reading of Scripture, among other practices but worship and singing as well. Church worship is to be demonstrated through voice and art. Worship practices would have mimicked those of the synagogue. According to the Talmud, there were 394 synagogues at the time of the destruction of the temple. Because the letter was written prior to the destruction of the temple, it can be reasonably argued that the worship practices Paul is referring to are those customs recently adopted by the early church.[14] Thus, actually performing psalms, hymns, and spiritual songs would have been a customary practice of the early church. Max Turner suggests, "Paul agonized over the well-being and spiritual growth of his congregations and felt personally responsible to ensure they were presented to Christ 'blameless' on the day of the Lord, lest his work be in vain"[15] and therefore instructed the church to use worship practices to teach and admonish one another as a way to guide the church in growing in the knowledge of Christ.[16] The development of theologically informed worship liturgy served to create the congregation's daily worship habits. Thus, Paul's exhortation to use psalms, hymns, and spiritual songs is an extension of his understanding of their usefulness within routine congregational practice.

COLOSSAE

The church at Colossae was not founded by Paul but by Epaphras, his "fellow servant,"[17] who was likely converted during Paul's three-year Ephesian ministry in approximately 52–55 CE. The text suggests that the believers

---

[14] There is disagreement as to the style and mount of vocalization and instrumentation during this time of transition. Instrumentation was limited post-temple destruction as a sign of mourning. Whether by 70 CE the NT church followed Jewish practice or stayed with what was already developed is unknown. Later writings, like the book of Revelation, which cite musical worship suggest that early Christians maintained instrumental worship at least until outlawed later in church history.

[15] Turner, "Spiritual Gifts," 189.

[16] In 1 Cor 9:19–23 Paul states that he will be all things and do all things in order that he may save some for the Lord. In this way, all means become necessary and sanctioned insofar as they are in line with the gospel message of Christ. Thus, the performance and utilization of psalms, hymns, and spiritual songs would seemingly be an acceptable method as well.

[17] Which explains its lack of mention in the book of Acts.

were converted gentiles and not converted Jews,[18] as demonstrated by the fact that there is no mention of the OT in the entire epistle. Written between 60 and 65 CE,[19] Paul did not personally visit Colossae prior to writing the letter. Epaphras was so concerned with the state of the young church that he visited Paul in prison—most likely in Rome[20]—further suggesting a dire situation arising in the church. Even though Epaphras visited Paul, Tychicus carried the letter to Colossae. Paul was concerned for them due to the syncretism that was being introduced to the church, leading to their reliance on practices other than faith in and supremacy of Christ. The letter is Paul's response to admonish the false teachers and ensure that the members of the church place their faith in Christ alone. It is not known how big the church at Colossae was, but at minimum there was one house church of Philemon, Apphia, and Archippus, and possibly a second at Nympha's home, suggesting a congregation of upwards of forty and possibly even up to one to two hundred. Nonetheless, Paul is thankful for them and what they have accomplished thus far.

Colossae is about one hundred miles from Ephesus and twelve miles from Laodicea and Hierapolis in Asia Minor in the Lycus River Valley. Under Roman control, Colossae belonged to the region of Phrygia. As was common in the Roman Empire, the city was multi-ethnic and of high mobility. Yet, not much is known of Colossae itself. It was located on two major trade routes between Ephesus and the Euphrates, both north-south and east-west. Colossae was once a thriving town of wool trade, but by the time of the letter it was in decline, with larger neighbors like Laodicea and Hierapolis gaining importance due to the major trade route moving twelve miles to pass through Laodicea. After an earthquake devastated the city in the early 60s, the city was rebuilt, however slowly, and never regained its prominence. To this day the remains at Colossae have not been excavated. It was mostly gentile by the mid-first century. According to Josephus, Colossae had a sizable migration of Jews who settled in the area in 213 BCE.[21] Being located on trade routes

---

[18] Moo ("Letters to the Colossians and to Philemon," 27–28) suggests the letter is focused solely on the gentiles because a majority of the things they are told to stop doing are typically gentile behaviors and not Jewish, with any Jewish behaviors easily being learned behaviors from early church practice.

[19] The exact years are debated; nevertheless, it was written pre-temple destruction and approximately one decade after the founding of the church.

[20] The letter is considered one of the "prison letters," most likely written during Paul's house arrest in Rome, though possibly in Caesarea. Because the epistle mentions being with Aristarchus, Paul's many workers, and his hopes to be released soon to come be with Philemon (which was written at the same time), house arrest in Rome fits best, even though Paul never distinctly states his location. When the author writes in a way that suggests the audience would make certain assumptions, the simplest explanation ought to be supported. In the case of location, as well as similarities to Ephesians and Philemon, Roman imprisonment makes the best case without added assumptions.

[21] According to Bruce (*Epistles to the Colossians*, 8): "The Phrygian inhabitants of the Lycus valley were only gradually Hellenized, except for those who lived in the cities. The new cities of Laodicea and Hierapolis were Greek cities from their foundation. When they came under Roman authority after 133 B.C.,

meant that it was a place where various philosophical views and religious practices would interact with one another, which explains the rise of the syncretistic religious movement that affected the church at Colossae. Yet, it also makes it difficult to decipher precisely to what issue Paul is writing.

EPHESUS

In contrast to Colossae, Ephesus was a major commercial center, located on the western coast of Asia Minor off the Aegean Sea, which made it the perfect central location for the establishment of the Pauline theology, from which neighboring churches would have been evangelized. Like Colossians, Ephesians is also a "prison letter."[22] Due to its similarity to Colossians, the conclusions can be drawn that both Colossians and Ephesians were penned together in Rome. The central themes, like Colossians, are unity of the church, the exaltation of Christ over one's self, and practical application of Christian living, both communally and within the family. F. F. Bruce calls Ephesians the summation of all Pauline writings and advances his teachings into a new stage.[23]

In the lead-up, Ephesians 5:18, Paul exhorts the congregation not to get "drunk on wine." Ephesus was in the middle of wine country, and over-indulgence was commonplace.[24] The Ephesians cultural practice included regular worship of the pagan wine-god Bacchus, which may have still been a common Christian practice in the church due to its widespread cultural usage in community festivals and holidays.[25] To worship their god, the pagans believed they had to be in a drunken state. Paul is thus setting the stage for the performance of psalms, hymns, and spiritual songs in the context of what is believed to be a raucous and celebratory society.

Even though considered to be written to the church at Ephesus, it is not as personal as other letters Paul wrote, especially considering he began his church planting ministry there. 3:1–7 makes it seem that the people are not fully aware of his ministry. Therefore, the letter was likely written to the whole of Asia Minor and was brought to Ephesus first by Tychicus who also

---

the cities were in some smaller degree Romanized, but none of them was reconstituted as a Roman colony, as several cities farther east were."

[22] Ephesians is understood to be the earliest of the NT books to be considered as Scripture.

[23] Bruce, *Epistles to the Colossians*, 229.

[24] In many modern evangelical churches, this verse is often used as a warning against drinking alcohol, and though there are merits to the thought itself—with many verses throughout the Bible that do compare drunkenness with foolery—in this instance, being "drunk on wine" is not a prohibition against alcohol consumption. Even if the conclusion could be made that it is saying one should not get drunk, that interpretation still falls short of the full intended meaning found here.

[25] In the same way, modern Christians may recognize Santa Claus as a festival figurehead, even though Christmas celebrations are about commemorating the birth of Jesus.

delivered Colossians. In fact, some ancient manuscripts do not contain the words "[in] Ephesus" (1:1) at all, further suggesting that this could have been intended for a broader regional audience.

Colossians cites a letter to the Laodiceans, which Marcion considered to be Ephesians.[26] However, Douglas Moo suggests that the letter to the Laodiceans is a now lost letter and not an alternate title for the Ephesians epistle.[27] Colossians and Ephesians are too alike not to be built upon one another. Which was written first is debatable, but Colossians in most cases is more concise while Ephesians develops Paul's arguments more fully.[28] This would suggest that Colossians came first, allowing Paul to build up his previous writing when penning Ephesians. Either way, the best understanding is that they were written at nearly the same time if not at the exact same time to be delivered simultaneously but for different reasons, with Paul demonstrating his genius and ability to write per the needs of the audience.

COLOSSIAN WORSHIP

The early church was a worshiping community, and therefore both the Colossians and Ephesians verses contain allusions to specific practices amid the backdrop of a Jewish-Greco-Roman world, as laid out in the two previous subsections. In *Letters 10.96*, Pliny the Younger (c.61–113 CE), governor of Bithynia, writes to Emperor Trajan asking what to do about the Christians. He includes a comment noting that the Christians would come together before dawn to "sing responsively a hymn to Christ as to a god." Tertullian (160–220 CE) noted the same practice, while adding that the Christians sang parts of the Holy Scripture. Likewise, Philo of Alexandria (c.25 BCE–50 CE) records the early Christians as a community in which the women and men lived together, sang in the context of mealtimes, and sang hymns both individual and communal. These descriptions exactly parallel the practices that can be gleaned from throughout the NT. In fact, writings throughout the Bible, Apocryphal New Testament books, extrabiblical sources, and writings from the church fathers contain hundreds of references to Christian singing and structured liturgical performance. As the date of authorship increases, the number of references likewise increases. This suggests that as the church grew both in number and confidence, structured liturgical worship became more ingrained into the church, even from an early date.

The Colossians lived in a pluralistic world of worship of multiple deities

---

[26] Utley, *Paul Bound*, 65.

[27] Moo, *Letters to the Colossians*, 26.

[28] Three Colossians passages share near exact phrasing and support for others found in Ephesians: 1:9b–25; 2:8–23; 3:14–16.

from Anatolian, Persian, Greek, Roman, Egyptians, and Jewish societies. The practices of each varied vastly, yet all believed that God/gods held a particular role in worship, whether as the object of worship or one from whom the worshiper would attempt to gain favor. With the rise of Hellenization, a robust multicultural exchange took place, including in musical influence. R. Kent Hughes reports the early church used songs in various ways: "one got up and sang perhaps from a psalm, and another answered antiphonally. Hymns broke forth in a heartfelt chorus. Others sang spontaneously about what God had done for them. There was music in their hearts."[29] In this way, Colossian worship would have been a collection of various religious practices and methods. "Music was intertwined with everyday life, and singing seems to have been integral to the emerging Christian community, with little sign of any negative attitude toward music."[30] Early Christian songs derived their content from the Scriptures and were used to promote the new church culture in regards to the progression of the Christian life.

Due to the persecution of Christians, worship in daily practice would likely be confined to the home, where the early churches developed. This may have meant a stronger influence on singing than instrumental worship, simply for the sake of volume and noise not because of an affliction against other methods.[31] Leland Ryken suggests music in the NT "is no longer priestly and professional. It is solidly social, congregational, and amateur."[32] Singing would have become more like speaking or a glorified chant. Some scholars suggest that musical instruments were not used at all because of the connection to both heathen activities and Jewish practice,[33] but there is absolutely no evidence of such a conclusion. Because the church was formed from congregants knowledgeable in gentile practices, their usual celebration methods would most likely have been integrated. At this early stage of development formalized ritual practices that new believers would ascribe to or exchange their regular daily habits for did not exist as is expected today. The Christians were themselves developing the practices for their own house churches that would only later become the stylized method for liturgical practices. Paul's warnings speak to practices found within gentile pagan ritual. Therefore, even in an early church learning a new liturgy that is founded within a Jewish or gentile realm would naturally also incorporate worship practices already familiar. Even though later church fathers of the fourth century pushed

---

[29] Hughes, *Colossians and Philemon*, 112.

[30] Begbie, *Resounding Truth*, 67–68.

[31] Church buildings were not present until the fourth century. Until that time, house churches were the worship centers for Christians. After Constantine in 313, Christianity could be openly practiced and probably overtook pagan rites; instruments were further allowed in the public arena. It was not until the political changes in the first few centuries CE that brought significant changes in the way worship would be done. Only then did the church fathers begin to have objections to the sounds of drums and symbols, instruments, and licentious singing.

[32] Ryken, *Liberated Imagination*, 51.

[33] Ryden, *Story of Our Hymns*, 16.

instrumentation out of practice, during the time of Paul's letter to the church at Colossae, it would likely have been their predisposition, especially with a young church as isolated as the Colossians.

While there is little knowledge of what truly happened in the early church, the text indicates that at minimum singing was a common practice and a significant part of the church gathering. Even if biblical scholars cannot find connections in religious texts, musicologists firmly believe music was most definitely a means of communication among the first-century communities.

> Both vocal and instrumental music played an integral role in festivals and religious ceremonies, theatrical productions, and rites of passage such as weddings and funerals. It was also a primary mode of formal and informal entertainment at meals, in homes, and while working.[34]

In fact, documentation for musical notation—even if not seen in our current biblical text—dates to the tenth century CE in Jewish culture and to the fifth century BCE in Greek culture. Musical notations served as memory aids to transmit and pass on belief systems through sound and memorization. Rudimentary examples can be found throughout the Psalter, with musical instruction and words of emphasis like *selah, lamnasseah,* and *miktam.*[35] Thus, just because the full meanings behind the notations in biblical documents did not survive to be interpreted into style and sound today does not mean they did not follow these in daily practice as per musicologists' findings from surrounding peoples and cultures.[36]

In conclusion, Paul quotes from Jewish Scriptures, writes in Greek, and addresses a congregation living in a Roman world. All three of these cultures have longstanding musical traditions that would have been clearly understood to the recipients of the letter.

---

[34] Hearon, "Music as a Medium," 180.

[35] The citations include: *lamnasseah* ("for the choirmaster"), *selah* (unknown musical direction), *higgayon* (unknown musical direction probably associated with whispers or meditation), *binginot/ binginot* ("stringed instruments"), *'el-hannehilot* ("with the flutes"), *'al-haggittit* ("according to the Gittith"), *'al-'alamot* ("according to the tune of Alamoth"), *'al-hasseminit* ("on the eighth"), *'al-tashet* ("Do Not Destroy"), *'al-mahalat* ("according to the tune of Makhalath"), *'al-mahalat le'annot* ("according to the tune of Makhalath Leanoth"), *'al-sosannim* ("according to the Lilies"), *'al-yonat 'elem rehoqim* ("to the Dove of Distant Oaks"), *'al-'ayyelet hassahar* ("to the Doe of the Dawn"), *'almut labben* ("to Death of the Son"), *sir* ("song"), *mizmor* ("a psalm"), *miktam* (a type of melody), *maskil* (something discerning and successful), *siggayon* (something wandering, possibly improvisation), *tehilla* ("a praise"), *tepilla* ("a prayer"), *hallelu yâ* ("praise to the Lord"), *ledawid* ("by David"), *lsslomo* ("by Solomon"), *leheman ha'ezrahi* ("by Heman the Ezrahite"), *le'etan ha'ezrahi* ("by Ethan the Ezrahite"), *lemoseh 'is-ha'elohim* ("by Moses the man of God"), *lidutun* ("by Jeduthun"), *libne-qorah* ("by/for the sons of Korah"), *sir-hanukkat habbayit* ("song for the dedication of the house"), *lehazkir* ("bring to remember" or "get God's attention"), *lelammed* ("for instruction"), *letoda, leyom hassabbat* ("for the day of Sabbath"), *hamma'alot* ("of ascents"), and *a-sap* ("Asaph"). For a complete list with explanations see Dale Brueggemann, "Psalms 4."

[36] Huldrych Zwingli referred to Ephesians and Colossians as considerations for how the church ought to incorporate music (Viladesau, *Theology and the Arts,* 213n14). At that time, even if antiphonal practice was the norm, just beginning to move into musical accompaniment, the Reformers believed that the early church was a musical, not solely vocal, people.

JEWISH CONTEXT

Paul was familiar with his audience even though he had never personally visited them. Presenting elements of Jewish worship with a gospel focus was important to Paul and served as a central pillar of the NT church in developing their own structured worship practices. Even though the Colossians were a gentile community in the Diaspora, at least in part, their worship practice would have mimicked common Jewish practices, which included: (1) no complex or rigid formality, though disorderly worship was also not permitted; (2) no indication that musical instruments were used, except maybe at certain rituals like funerals after the destruction of the temple;[37] and (3) no evident exclusion of women from singing.[38] Prayers and praise were central aspects of synagogue worship, from which derived the Jewish creed, the Shema, the confession of faith and benediction. Their practices included a reading of the Law and Prophets, a discourse, prayer, the priestly blessing, the amen, and singing of the Psalter. The epistle varies drastically from Paul's instinct to present Christ as the promised Jewish Messiah and fulfillment of OT prophecy. Nevertheless, many themes present in the epistle parallel those found in the Jewish Scriptures and served as fundamental means for the development of the early church. In fact, Paul's admonitions against lying, stealing, and drunkenness were prevalent themes in Second Temple Judaism and Jewish Christianity.

Instruction was a key part of both temple and synagogue practice, which included a reading of the Scriptures, the exposition of a homily of instructive living, and an application to their lives. Jewish tradition changed drastically pre- and post-temple destruction in 70 CE.[39] The Jewish culture used music vastly as part of their teaching during the Second Temple period, while post-temple it is thought that the Pharisees did not sanction performance worship due to instituting a time of mourning. Worship practices, therefore, varied from synagogue to synagogue once the temple no longer served as the central model for Israelite worship.[40] One common practice was the chanting of old

---

[37] It is important to remember that the penning of the letter came prior to the destruction of the temple. Even though Jewish practices changed after its destruction, the understanding by Paul at the time of writing most likely would have included instrumentation because the temple musicians were known to play. This would have then also carried over to everyday practice.

[38] Janzen, *Psalms, Hymns and Spiritual Songs*, 1022–28; Smith, "Ancient Synagogue," 15.

[39] After the destruction of the temple in 70 CE by Titus's army, primary worship shifted from the temple to the synagogue and home. This transition meant that worship would become less centralized and for NT Christians developed without a model for common practice.

[40] For the Colossians, this allowed Greek religious practices to enter the worship liturgy, even though it may have still been based on temple practices. The lack of a central model meant that the worship could be molded to fit the local congregation as well as incorporate known practices of the geographic area and societal norms.

psalms, which were understood Christologically.[41]

> In Old Testament times [pre-destruction], the form of praise was more organized and dramatized than what we find in the early days of the church. . . . [For NT Christians,] singing seems to have been characterized more by spontaneity, simplicity, and sincerity.[42]

This would have included instrumental accompaniment to establish the tone of the worship, whether celebratory, lament, or otherwise.[43]

As a former Jew, Paul would most likely have had customary Jewish liturgical practices in mind even while writing a converted gentile congregation, including the most common: alteration, responsorial psalmody, antiphonal psalmody, solo singing of Scripture with an "amen" response, musical instrumentation of various drumming devices, blown trumpets and similar instruments, symbols and tambourines, and stringed instruments like the lyre and harp. The addition of music added a spiritual element to the vocal aspect. Levitical practices in the temple included instrumentation and formal singing and it is commonly thought that this influenced both the synagogue and early Christian church. However, there is no formal evidence of the such. Per the arguments of musicologists, they likely would have formed traditions according to practices with which they were accustomed. For example, the Jews cherished the messianic psalms; the singing of psalms gave rise to antiphonal singing between minister and congregation, and the psalms and hymnody became the first known worship practices of the early church, which compels an a priori understanding.

It is also commonly thought that musical instruments were banned from the early church due to their worldly nature and to lament the destruction of the temple, just as they were banned after the destruction of the temple within Jewish circles. However, findings suggest there is absolutely no support for the idea that instruments were banned at all.[44] Likewise, there is no evidence either way of the supposed ban on musical instruments after the temple destruction. Most likely, instrumentation in Jewish practice was used for certain occasions but probably not sanctioned religious practices. "The idea of a legalistic rabbinical ban on the use of musical instruments is a piece of latter-day etiology without historical basis."[45] With the temple destroyed, Jews

---

[41] As noted above and in the previous chapter in regards to Hebrews 2:12–13.

[42] Olford and Olford, "Preacher and Music," 323. It is important to look at how "singing" was meant in context. Throughout both the OT and NT, the words "speak" and "sing" have added connotations beyond merely speaking and singing. A modern example: If some says the phrase "worship band" in Western culture today, most people assume a combination of a standard drum set, electric guitar, acoustic guitar, bass guitar, keyboard, and vocalists. However, the ingredients of a "band" will vary over time depending upon musical tastes, style, and audience. Therefore, in regards to the text, to say "singing" can also mean playing instruments without much concern as to whether it was understood as the common practice. I expect this is the case in this instance.

[43] Note musical notations found throughout the Psalter, for example.

[44] Porter, "Music," 713.

[45] Smith, "Ancient Synagogue," 3.

would attend synagogues, and would eventually establish their own individual contextual understanding for worship. Because the synagogue did not have professional musicians, instrumentation would likely have become more informal.

Last, there is a critical aspect overlooked in every commentary on Colossians that I explored: every one presumed an understanding from a post-temple destruction to describe the worship practices Paul was writing about,[46] even though he would have written the letter pre-destruction. The Colossian society was far removed from much temple influence in the first place and probably only referred to it as a general model. For scholars to present worship practices from a viewpoint inconceivable to the Colossian people—or any Jew or Christian for that matter—is highly problematic. Every scholar is guilty of writing a history back into the letter that could not have even been imagined by the original audience. A Pharisee—like Paul previously was—post-destruction would have presumably been against music, yet Paul puts it in high regard, as seen by placing it in the "put-on" section. It is what the church is to do. Therefore, the text is best understood with psalms, hymns, and spiritual songs having an inherently positive production significance within structured liturgical practices.

GRECO-ROMAN CONTEXT

Paul was from Tarsus, one of the major religious and cultural centers of Greek civilization in the first century, which would make him familiar with musical practices in Hellenistic society. The Greco-Roman musical culture had a large influence on early gentile Christians, with both musical and liturgical practices of the Hellenistic world taking root in early Christian practice, especially in churches founded within the Pauline sphere of Asia Minor. The Greeks believed music was of divine origin and therefore attributed musical qualities to their various gods. Likewise, they believed music possessed higher powers that could be used for healing and purifying the mind and body, working miracles, and influencing human thought.[47] Roman civilizations would transmit cultural memories through the rhythm and melody of song. Hymns became praises to the gods by focusing on the acts they performed. Sounds were thought to hold ethical and moral effects in that they could

---

[46] Possibly in an inadvertent attempt to minimize the importance of worship in order to keep preaching central?

[47] Philo in Therapeutae notes the people's use of hands and feet to create a choralistic dance of keeping time and engaging with the music. By adding movement to the rhythm, the ability to teach and remember the material increases because the actions and rhythmic themes become associated with the material. Even though later writers criticized the practice, Theodoret of Cyrus contends that the practice did exist and probably was not of concern at the time of the early Jesus communities.

affect the will. For this reason, in Greco-Roman society music and song were likewise believed to have the ability to teach. Shrines to the gods were commonplace throughout the ANE, where citizens would perform hymns, prayers, sacrifices, and festivals of song, dance, wine, and food associated with them, because they believed that the gods held the central order of all things from blessing to the cycle of nature, to good harvest, to diseases, and to favor in warfare.

As the church incorporated Greco-Roman practices, practices performed to pagan gods would likely have been integrated yet incorporated as Christian hymnody. In Greek tradition, melody and poetry were closely related; music most always contained words, while poetry was set to music and was metered and rhythmic. Musical performances were mostly improvised by a skilled performer performing hymnos, lofty art songs in praise of the gods and heroic songs that were thought to possess divine inspiration. In Christian communities of the Greco-Roman world, there was no differentiation based on age, sex, race, or class when it came to worship practice, and importantly, most liturgies were unchallenged by occupying forces. Anyone in the community could participate in the whole: literally, one another could influence one another.

Accordingly, the Roman writings of Cicero and Quintilian show that there was a strong familiarity with music. Large choruses and orchestras, grand musical festivals and competitions, and songs essential to every social activity were all incorporated. Themes present in their music included work, recreation, god worship, festivals, holiday celebrations, satires, love songs, and drinking songs. Many emperors were considered musical patrons, with even Nero gaining personal fame for his musical ability. Roman music was mostly created for military purposes and relied heavily on the brass instruments of trumpets and horns. Conventional instruments included the lyre and aulos, one being calm and one exciting, signifying a tension between higher beauty and uninhibited behavior.

Paul's writing, likewise, has roots in the local culture, often drawing imagery from local practices and architecture (Acts 17:16–34). Pagan cultures separated morality from religion and worshiped through intoxicating and heathen festivities. Yet, Christian worship was to abandon carnal activities and be filled with the Spirit. In this way, the early church most likely struggled to separate itself from common cultural activities found in everyday festivals, which explains the warnings within the "put-off" section. The Greek and Roman worship practices included a mass emotional ecstasy dedicated to Dionysus and Bacchus. Drunkenness was thought to bring them to a higher state of being, which in turn connected them the gods. In religious practice the people would fill themselves with wine and then spurt out utterances thought to be of divine inspiration.[48] When Christians incorporated both pagan and Jewish practices, while removing the need for excessive consumption in order to commune with the Christ, charismatic singing could be viewed as a natural addition.[49]

AN ALTERNATIVE UNDERSTANDING

The Greeks, Romans, and Jews all participated in worship practices as defined by their social loci. It is, therefore, important not to automatically presume Paul's citation of psalms, hymns, and spiritual songs can be understood from any one worldview. Contrary to the common understanding, the worship context of Colossians 3:16—if not the book of Colossians in its entirety—ought to therefore be viewed as much from a Greco-Roman viewpoint as a Jewish one. Paul never alters the audience,[50] demonstrating that he understands that his recipients come from one particular cultural landscape. However, in Ephesians Paul develops the Colossian arguments with Jewish citations and connotations, showing that it built upon the Colossian letter for use with a larger audience. The Greeks would not have necessarily viewed psalms, hymns, and spiritual songs with a Jewish religious connotation, but rather would do so through their social context. Thus, Paul takes pagan understandings and weaves in themes supporting the supremacy of Christ. It is important, therefore, not to make the mistake of writing Jewish history back into Paul's writing in this case. Nevertheless, even though it is written to Greco-Roman Christians with specific worship practices, Jewish undertones are still present and applicable to Jewish-Christians as seen through the Ephesians passage.

So, which is it, Greek, Roman, Jewish, or pagan? Yes. The worship themes presented are a collaboration of all first-century belief systems that influenced early Christianity. This is important and relevant to the greater understanding of church technical arts because of the modern desire to "nail down" a right and wrong way to present worship. Should a lyrics slide be limited to two lines? Maybe four? Should lights be on during worship? Or dimmed? Where certain practices will "speak" to a particular demographic better than others, first-century worship practice shows that all traditions work together to inform praxis.

---

[48] Viljoen, "Song and Music," 438–39.

[49] In Acts 2:13–18 the apostles were thought to be drunk when the Spirit came upon them. Their actions included outbursts of speaking in tongues—or languages common to the location of future evangelism but not previously known to the disciple. Even though the activity is different, the societal assumption demonstrates that outward drunken religious expressions were common.

[50] The epistle, written to converted gentiles, contains no direct Old Testament citations, with only possible allusions to Jewish themes.

THE COLOSSIAN HERESY

It is believed that Paul's entire purpose for writing the letter is to fight a false teaching known as the "Colossian Heresy." The Colossians were faithful, yet there was an attractive but dangerous teaching being introduced to the congregation. Paul wrote in response to the urgent need. Even though it does not affect the meaning of Colossian worship, knowing the worldview Paul is writing against can help to place the produced performance methods of psalms, hymns, and spiritual songs into perspective as a means of teaching and admonishing.

Paul does not explicitly name the false teachers nor the false teaching itself. The letter focuses on the aspects of the gospel that the heresy threatened. The Colossian Heresy promoted some type of asceticism. It should not be thought of as a new way of understanding the gospel, i.e., what modern Christians think of as a false Bible teacher, but rather the thoughts of the world and modern culture. It included various practices that would have been known: philosophy and intellectualizing religion (2:8), human tradition (2:8, 22), elemental spiritual forces of the world (2:8), lack of focus on Christ (2:8), circumcision (2:10), baptism (2:11), Jewish food restrictions and holy days (2:16), ascetic practices (2:18, 23), angel worship (2:18), visions (2:18), pride from unspiritual minds (2:18), not recognizing Christ as the head of the church (2:19), and using worldly rules as a means to spiritual growth (2:20–23). This teaching arose either from within the church itself and not from an outside force, or from the congregation bringing in other ideas to their already established church beliefs.[51] Viewing the epistle in this way helps to explain why Paul never discusses anything outside the church corpus. Keeping the passage Christocentric, it aids Paul's argument against Gnosticism: the solution is a firm grip on the Christology of the gospel. In structured liturgical terms, what is in view is a gentile church fighting with former pagan worship practices while being instructed on proper Jewish worship methods. Paul seeks to place these into proper perspective by making Christ the sole focus of their worship, both individually and corporately.

MUSICAL METHODOLOGY:
PSALMS, HYMNS, AND SPIRITUAL SONGS?

Paul had a propensity for inserting hymnal and creedal pieces throughout

---

[51] Moo, *Letters to the Colossians*, 47.

144

his epistles. The high point and opening to Colossians is the hymn found in 1:15–20 about the exalted position of Christ.[52] Even there Paul himself is teaching through literary-musicology.[53] What then are we to make of the phrase "psalms, hymns, and spiritual songs?" It is not definite whether the three terms are a heaping-together of synonyms, a stylistic feature, different aspects that inventory early Christian hymnody within church practice, the OT psalms, liturgical methodology, or spontaneous Christian song. Nevertheless, the verse sets apart musicians and church music as something special to the edification of the corporate body. There are two main schools of thought: psalms, hymns, and spiritual songs were either synonymous terms or were entirely individual words representing distinctly different worship techniques.

To some modern scholars the terms cannot be entirely deciphered as fully distinct because they all are used interchangeably for each other throughout Scripture, with all three referring to the Psalms. Many suggest there can be no hard demarcation in the three terms, even if they do have varying allusions to a wide range of singing and praise practices.[54] Douglas Moo goes so far as to say that where we can distinguish meaning from the three phrases, the differences are questionable.[55] Nevertheless, even if the three words seem linguistically synonymous, they still represent the beginnings of church liturgy. Grant Osborne and Jeremy Begbie move even further away, to say psalms, hymns, and spiritual songs were most definitely not intended to be viewed as separate aspects of worship music, and though people want to draw distinctions between the three phrases, we cannot.[56] However, no observations we can make should be dismissed.

Other scholars believe that even though psalms, hymns, and spiritual songs have notable overlap, significant distinctions can indeed be made. D. A. Carson suggests the terms are not synonymous because they fall within a section noting a very specific list of the workings that those who are filled with the Spirit perform, like teaching, admonishing, giving thanks, and praising.[57] Since it would be improper to group together other workings of the Spirit, it would be improper to do so here as well. Listing them individually denotes that each one is a specific type of musical song for use in the edification of the congregation. In the gentile-Christian world the three terms

---

[52] There is disagreement as to whether Col 1:15–20 is a preexisting hymn or one penned by Paul in the course of authoring the epistle. Either option does not impact the argument here, and still demonstrates that hymnody was a literary device employed by him.

[53] I believe it is important to note that no music survives from late antiquity, so how the hymnic writings were to be presented is unknown outside of any natural alliteration or rhythmic countenance that may exist in the original languages. However, the musical notations found within the text and in other archaeological finds confirms that music for teaching and auditory style did exist in worship practice.

[54] Crabtree, "Colossians 3:16," 12; Fowl, *Ephesians*, 177.

[55] Moo, *Letters to the Colossians*, 289.

[56] Osborne, *Hermeneutical Spiral*, 113.; Begbie, *Resounding Truth*, 69–70.

[57] Carson, *Worship*, 118.

were distinct, while also serving as "the three most important terms used in the Septuagint to describe a religious song, and all refer to believers' songs of praise to God."[58] Mid-twentieth-century musicologist Egon Wellesz finds that psalms, hymns, and spiritual songs were thought to be three distinct items referring to synagogue psalmody, syllabic songs of praise of paraphrased biblical texts, and chants of ecstatic nature.[59] His classifications have formed the basic understanding for those who believe that each one has a specific meaning and use in worship practice.

Either way, the contemporary disagreement over musical preference can be answered in this verse: with three different types of worship liturgy mentioned, Paul appears to make an allowance for various forms of worship. In this way, all worship practices today—including the technical arts—would then be possible and sanctioned assuming they are not in conflict with other commandments of Scripture.

PSALMS

> Evidence of psalmody in the Christian church is cited from writers in the first centuries outside of the Jewish tradition who describe [it] as an unusual form of music. . . . Writers from within the Jewish tradition find nothing exceptional about it, which suggests that it is familiar to them.[60]

For this reason, most scholars consider Paul's notation of "psalms" to refer to the OT psalms. They are thought to have served as the chief vehicle for Christian praise from early on, shaping the foundation of early Christian doxology. The psalms cover nine distinct genres, all emphasizing personal expression from a Jewish perspective: praise, lament, thanksgiving, confidence, remembrance, wisdom, kingship, petition, and prophecy. The Israelites even referred to the book of Psalms as the "book of Praises;" *psallb* means to sing a song of praise, which was the singing of a spiritual or sacred song. Early Christians called it the "Psalter" or "poems set to music." In this way, they are not simply to be recited, but performed, lending importance to the methods used to create the experience. In the Greco-Roman culture a psalm was understood to be a praise song, which could either be set to music or stand on its own. Of the seven NT uses of *psalmos*[61] all of them outside of Paul's use here definitively refer to the Psalter.

For the Jews, there may have been an eight-hundred-year dead period

---

[58] Hoehner, *Ephesians*, 14170–71.
[59] Smith, "Ancient Synagogue," 1–3.
[60] Porter, "Music," 713.
[61] Luke 20:42; 24:44; Acts 1:20; 13:33; 1 Cor 14:26; Eph 5:19; and Col 3:16.

where worship was primarily prayers and readings, with instrumentation only taking form after the exile and ceased again after temple-destruction. Christians, however, are thought to have sung from the beginning in their worship practices, which then continued to develop over time. If Paul is referring to taking psalms directly from Scripture, this could also include being set to the sounds of a skilled musician as directed by the psalms' authors, mimicking the practice performed at the temple by the dedicated temple artisans.[62] As well, the NT Christians focused on very different psalms from the Jews, viewing the Psalter as messianic, pointing toward Christ as the promised Savior.[63] Thus, their performance of them most likely became more personalized as well. James Janzen summarizes the purpose of the psalms:

> (A) personal experiential viewpoint; (B) personal expression of life's highs and lows combined with petitions to God; (C) focus on God is subjective, emphasizing personal relationship and experience; (D) song(s) that possess characteristics of one or more psalm type(s); (E) musical accompaniment, literally the act of playing an instrument; (F) part of Hebrew culture.[64]

Psalmos literally means "plucking" the string of an instrument or bowl. In its Greek connotation, it is a musical instrumental notation and not vocal or poetic as in the Jewish context. To a Hellenized recipient of Paul's letter, it can be reasonably proposed that instrumentation was indeed supplemental to vocal worship. This understanding could conceivably have been the accepted view of the congregation and is paralleled by the written musical notations and conductor-like directions to the worship leaders found atop the Psalter. In other words, an instrumental culture would be naturally inclined to offer musical meaning to existing musical notations, even if not the same notations as intended by the original authors or if the exact musical understanding had been lost by the first century CE. Still, the fact remains that the notations are present and suggests they were therefore followed by a musically inclined culture. Thus, psalms in the Greco-Roman world would be songs of both vocal singing and instrumental accompaniment.

The purpose of the Psalter was thought to create calmness and harmony. The psalms were:

> Poetry, which had certain characteristics: (1) filled with stress and accent; (2) its metre was regular occurrence of longs and shorts in fixed ratios; (3) it had poetic parallelism which served as a form of repetition; (4) strophic structure is thought of as the verse, and when a parallel of the thought is presented, it comes next; (5) it has poetic assonance of repetition of the same vowel; (6) possessed alliteration; and (7) has

---

[62] I suggest that it is important to note that even though this hypothesis appears to have substantial support, it is still conjecture configured from circumstantial evidence. Nevertheless, it is the view held here.
[63] Like Ps 22, from Heb 2:12.
[64] Janzen, *Psalms, Hymns and Spiritual Songs*, 918–32.

poetic rhyme.[65]

A core element of psalms is that they are didactic in nature and, therefore, to use them for the purpose of teaching and admonishing makes for the best understanding of the verse. This is precisely why the NT writers quote the psalms often, as seen in Romans, Hebrews, and Acts.[66] The psalms possess doxological, hortatory, and edifying components: they instruct and teach as they are performed, preached, taught, and meditated upon regularly. Historically, OT songs were often used for teaching and reminding the Jews of the identity of YHWH and his deeds. Psalms, therefore, are well suited for instruction and admonition. This possibly explains why Paul would instruct a Greek Christian congregation to add psalms into their established musical practice.

HYMNS

> The hymn . . . is an ode or poem addressed to, and in honor, praise, or adoration of, someone or something [usually a divine entity]. . . . The Christian hymn is a text addressed to, and in honor, praise, or adoration of God, as the Triune Godhead, or to any single member of the Trinity.[67]

Hymns are like psalms. However, where psalms carry various meanings, hymns must denote praise, regardless of any formalized style. A hymn can be a vocal or instrumental song of praise, frequently with stringed accompaniment. James Janzen summarizes the key ideas found in hymns: "(1) a song in praise of gods, heroes, and conquerors; (2) extolling the virtues and characteristics of God; (3) eloquent and artful weaving of text; [and] (4) a sacred song of praise to the God of the universe."[68] They were directed toward God as a sign of devotion and appeal.

In Greco-Roman society, hymns were festive songs in celebration to deities, used to show reverence and respect; the gods were thought to be active and able to bring about blessings, like good crops and health, when sung as a devotion to them. In Christian practice, the hymn is poetry representing the Christian life focused on Christ, the aspects of God's divine activity, and the application to daily activities. Hymns were the focus of the early church and were the first steps in creating original Christian worship material. It is uncommon to find petition in hymns. Most focus on praise and present beliefs

---

[65] Sasser, "Worship," 101–2.

[66] Another reason for the authors' many citations is because Christ himself quoted them and not only because they are didactic. Thus, because Jesus viewed them as didactic and good for teaching and admonishing, so too would the NT authors.

[67] Sasser, "Worship," 10.

[68] Janzen, *Psalms, Hymns and Spiritual Songs*, 850–54.

of the Christian faith as opposed to other liturgical practices like prayer, where petition is common.[69]

The word "hymn" is found six times in the NT. The only uses of "hymn" as a noun are found in Colossians 3:16 and Ephesians 5:19. Its use here may be as a "festive hymn of praise."[70] This is a distinct possibility when viewed within the context of the joyous and boisterous Greco-Roman worship of Bacchus. As a verb, hymn ("to sing a hymn") occurs four times in the NT: Matthew 26:30; Mark 14:26; Acts 16:25; and Hebrews 2:12. In Matthew and Mark, the term is most certainly referring to an OT psalm; Jesus and his disciples are partaking in the Passover meal and the common liturgical practices associated with it, likely singing a traditional Hallel song. In Acts, Paul and Silas are corporately singing and praying in prison; the usage here is unknown, but most likely it was a Christian adaptation of praise. In Hebrews 2:12 the use is a citation of Psalm 22:22, a messianic psalm about Christ himself.

Other known hymns in Paul's letters include Philippians 2:5–11; Colossians 1:15–20; and 1 Timothy 3:16, though there is no denotation of *hymnos* introducing them. This should not be overlooked. The inclusion of hymns within his hortatory texts demonstrates that Paul was himself a worship artist.[71] Even though not referred to as hymns, Revelation 4 and 5 are the type of hymns that might have been sung in the early church. In Revelation, the playing of instruments is displayed alongside vocal singing. *Hymnos* appears seventy-one times in the LXX, referring to religious songs, especially songs of praise to God.

Finally, hymns can be viewed as either the musical deliverance of a biblical text or canticles, Christian writings as opposed to psalms which would have Jewish history attached. They were new songs with expressions of Christology. Musically, hymns are syllabic in nature, with simple melodies and syllables carried across one or two notes. In Hellenistic society, hymns were performed and written with high levels of craftsmanship, a mark of the Greco-Roman world. Hymns served as the principal vehicle for passing down historical events from generation to generation. Hymns incorporated sharing cultural truth and represented a form of art for oral cultures and, therefore, had to integrate artistic literary and musical devices in order to be effective. Where hymns were semi-common in Jewish writings, they were presumably

---

[69] Martin ("Worship," 420) refers to a hymn as a "specifically Christian genre" with devotion to the Lord. He suggests it a Christ psalm. This is problematic in a strict sense, because hymns are indeed found in cultures outside of the Christian loci and are used in the Gospels to refer to the OT Psalter, but there is no doubt that early Christians began creating their own worship books based on the activities and teachings of Christ, which believers today consider "hymns." Whereas the Jews relied on established Scripture, Christians borrowed, changed, developed, and refocused Scripture to fit their understanding of Christ.

[70] Melick, *Philippians, Colossians, Philemon*, 305.

[71] Applying this to the focus of this book, Paul's use can likewise be viewed to demonstrate that musical artistry through the presentation of the biblical text is a valid method for teaching.

considered the norm in Greek religious circles and likely would have been well known to the audience to whom Paul wrote.[72]

SPIRITUAL SONGS

Colossians 3:16 and Ephesians 5:19 are the only two verses in which Paul uses the word *ode* and in doing so adds the modifier "spiritual." Spiritual could modify all three words—as some scholars argue—however, it grammatically better refers to the last noun, song/ode, specifically adding a religious connotation to it. It suggests a spontaneous, charismatic, and ecstatic church hymnody in the context of congregational singing.[73] "Spiritual" could refer to either the content of the song itself, being the work of one filled with the Spirit, or a reflection of the person of Christ. Whichever it refers to, Paul expects odes to be spiritual in nature, not frivolous or trivial; they are to represent a person changed by the saving grace of Christ, one who has put-on the new self. "Spiritual songs" connects well to Paul's main argument of the epistle. James Janzen summarizes its role:

> Under the guidance of the Holy Spirit, the spiritual song was a distinct genre not primarily concerned with the praise, petition, lament, and thanksgiving of the psalms or the poetic doctrinal focus of hymns. Instead, this genre was primarily an outburst of praise and joy, where text played a secondary role. The spiritual song provided a medium of response. . . . It provided an emotional release for the surge of feelings that rise from the powerful work of "the Word of Christ dwelling richly." . . . The spiritual song facilitates the infilling and control of the Holy Spirit, allowing Him to do His work of teaching, guiding, overcoming, interceding, controlling, comforting, convicting, regenerating, and sanctifying.[74]

Ode is the general Greek word for song or ode (English) and can refer to any song. In Greek tragedy, odes are songs of joy, praise, poems, or singing and can have music, dancing, or rejoicing accompanying it. Traditionally, heroes were thought to have sung these personally in public forums and banquets. Odes were commonly improvised and thought to have divine inspiration. Paul probably anticipated a similar inspiration, but inspiration from the

---

[72] Tertullian noted that at the *agape* meal, each person would stand and sing a hymn to God, either from Scripture or of his own composing. This demonstrates that there was a certain level of poetic license allowed in the presentation and composition of hymns. Creativity would have been a central aspect of presenting truth.

[73] Detwiler ("Church Music and Colossians 3:16, 362) suggests that the charismatic sense is most likely not the act of speaking in tongues because teaching and admonishing need to be easily understood. Likewise, Begbie (*Resounding Truth*, 71) contends spiritual songs is not a stirring in the sense of speaking in tongues because it is specifically expected to have a purpose, the purpose of building up one another.

[74] Janzen, *Psalms, Hymns and Spiritual Songs*, 1332–44.

Holy Spirit. There would be room for expression of on-the-spot gesturing that is lively, captivating, and impulsive.

Even though *ode* can refer to a general song, in the Bible an *ode* is always sung as a praise to God without exception, even if the content is exhorting or pertaining to particular matters of interest. Paul appears to specify that religious emphasis must be present in the content; however, the style could be any type familiar to the congregation. In Greek culture, *ode* typically was a poem that was sung. In this way, "spiritual songs" can be understood as musical compositions similar to modern gospel songs—those which utilize contemporary, spontaneous outbursts stylistically understood by that particular audience—with content that lifts praise to God. They are religiously themed and not secular. In practice, "They are 'spiritual' in that the Holy Spirit is viewed as actively inspiring the composers as they write their songs of praise to the risen Christ and what he has accomplished by his work on the cross."[75]

The Greek phrase for "spiritual songs" is *odaes pneumaticaes*, may possibly be best understood as "odes upon the breath," since the word *pneuma* has a dual meaning of both "breath" and "spirit." "Derivatives have the following meanings: *pneo*, to blow (of wind and air generally, also on a musical instrument); to breathe (also in the sense of to be alive); to emit a fragrance, etc.; to radiate heat, anger, courage, benevolence, etc."[76] There are parallel uses throughout the Bible.[77] Just as the Spirit and breath (Gen 2:6–7) find parallels throughout the creation narrative, it is fitting that Paul would use a word with dual meaning for an act that is a literal pushing out of breath and sound by those filled with the Spirit of God. *Pneuma* is a word with special meaning for NT church. It is the Spirit of God who works in all believers, and thus the power that is to guide song and instruction.

Outside of Colossians 3:16 and Ephesians 5:19, *ode* only appears in the NT five times, all in Revelation: it is used as a modifier in 5:9, in 14:3 twice as "new song," and in 15:3 as "song of Moses" and "song of the Lamb." Throughout the Bible several types of songs/odes are encompassed for many different occasions, including weddings, working, love songs, funeral songs, laments, war songs, praises, drinking songs, prayers, alleluias, chants, and sacrificial songs. Many NT passages may, in fact, be odes due to their apparent rhythmic and lyrical composition. Some are well known, including the Magnificat (Luke 1:46–55); song of Zechariah (Luke 1:67–79); song of the angels (Luke 2:14); 1 Timothy 3:16; Ephesians 5:14; Philippians 2:6–11; and Revelation 4:8, 11; 5:9–10; 11:17–18; 15:3–4; and 19:1–8. There may be others scholars are unaware of because they can only be viewed based on a contemporary understanding of song composition, and the modern reader cannot

---

[75] Arnold, *Ephesians*, 354.
[76] Janzen, *Psalms, Hymns and Spiritual Songs*, 873–75.
[77] Gen 6:7 LXX; Isa 11:4 LXX; 2 Thess 2:8; and Rev 11:11.

interpret them being two thousand years removed.[78]

Songs are sermons in themselves when flowing out of a heart that has richly dwelled on the Word of God. Even though there is no historical proof, "spiritual songs" likely gave rise to "symphonal" communal singing because as a special musical type it would engage the audience in a way that would move them, not being tied to written Scripture of hortatory functionality.[79] The corporate focus would provide a theological vehicle good for use by early Christian serving as lyrical outpourings of the developing church.[80] The informality of "spiritual songs" does not diminish their impact: the Spirit's work makes their meaning more powerful and offers freedom in delivery for the good of the congregation by and through teaching and admonishing.

CONCLUSIONS ON PHRASEOLOGY

Paul is not attempting to be redundant in penning the phrase "psalms, hymns, and spiritual songs"—even though a string of synonymous terms is a common literary style of his—but rather appears to be showing that there is a vast richness that comes through utilizing a variety of worship elements when performed corporately. The use of three different terms is significant. Though it is uncertain if clear distinctions should be made since the terms are used interchangeably throughout the Bible, the evidence demonstrates that they most definitely should not be considered wholly synonymous, specifically because they held different meanings depending upon the cultural loci. Have traits shared among them does not negate distinctive identities. There could also exist other meanings than these presented here that cannot be recognized because the musical style did not survive. Likewise, even though three terms are listed—accounting for the distinct Greco-Roman and Jewish contexts—there are a minimum of six different possible understandings of the terms.[81] Therefore, it is less important to strictly define each one

---

[78] This is obviously fully circumstantial and presumptive, yet highly probable nonetheless. Just as the Psalter contains obvious musical notations that today have lost their original meaning, it is highly likely that NT writings that are literary works of art in themselves would contain other instances of musical connotations that have been lost. Just as this book is doing, pulling out worship themes from texts not normally understood as such, there are likely others which have been lost over time.

[79] Lange, *Ephesians*, 194.

[80] Some scholars have argued that gospel songs—modern spiritual songs—are not biblically founded, and the church should return to hymn singing. This is problematic because it rejects the artistic nature of song that the *ode* allows for. This position loses focus on Paul's call to utilize all methods for teaching and admonishing through song. It does not need to be tied to Scripture. The text here suggests that the Spirit can move in ways that are still deemed biblical.

[81] It is commonly thought that the Jesus communities took the practices of the Jewish synagogue even though no proof of any singing is noted in any sources. Yet, Paul in 1 Cor 14 specifically notes the early church sang. Because they did, maybe it was not the synagogue from where the practices developed. The Greco-Roman gentiles would have understood these terms to possess their own distinct meanings. They

than it is to recognize that Paul is recommending a broader variety of worship styles, mediums, and practices, insofar as each still teaches and admonishes the corporate body.

There is no specific textual proof from the epistle alone that the Colossians utilized instrumentation, though the verse is open to it and the historical knowledge of the societal norms suggests that there is a high probability of it. In the same way, there is no clear appeal against instrumentation in worship. While "it seems to have been a foregone conclusion among early Christian that instruments had no place in public worship,"[82] these findings suggest this conclusion is an unfounded back-writing of church history into first-century-gentile practices. Produced worship practices not only existed, but were the common practice. Before the destruction of the temple, there was no restriction on instrumentation, which is when early Christianity would likely have begun to form its worship and structured liturgical foundation. Even if restricted to house churches for fear of persecution, there was no policy against multi-medium practices. Psalms, hymns, and spiritual songs all have an ingrained holistic performance aspect. *Psalmos*, for example, suggests the use of musical instruments, laments, and songs of praise.

It is interesting that in the scholarly literature, even those scholars who claim that psalms, hymns, and spiritual songs should not be viewed with individual meanings often continue to define them separately within their commentaries. By doing so, they further remove the phrase from its greater context, defeating their objective of considering them as one unit. The examination of various meanings among the three shows that Paul was deliberate in his selection of terms, allowing him to properly speak to a Gentile church faced with differing philosophies of how to follow their new-found religion. He incorporated a Greco-Roman cultural understanding into a Jewish-based religion. In this way, it is fully possible the Colossian Heresy was confusing the understanding of how pagan ritual could and should shape itself into Jewish-Christianity.

Paul's actions were intentional and specifically chosen to form a holistic view of produced worship in praxis. It is too simplistic to say "psalms" are the OT Psalter, "hymns" are songs about Christ, and "spiritual songs" are spontaneous compositions of the Holy Spirit. They certainly are those things, but they are much more, incorporating both traditional Jewish and pagan Greco-Roman meanings. In this way, Paul could develop the text from the Colossian usage for the Ephesians, recalling the same terms but with a meaning specific to their church. Biblical musicologists, as opposed to strict textual scholars, presume the terms signify multi-medium music being made.

---

are better understood as a collection of worship/musical styles collected from all surrounding communities and cultural influences.
[82] Brown, "Music," 6307.

Therefore, the best way to understand early Christian worship is not as an either/or but a both/and proposition. Paul redefined liturgical worship for the Christian church by blending together regional Greco-Roman ritual with Jewish historical methods. Each offers something to the other. Greeks offer music and celebration; the Jews offer tradition and messianic meaning. In this way, Paul demonstrates that there is more to Christian worship than the Psalter, as commonly thought. New Testament worship involves a mixture of meaning, methodology, and practice. For the modern church technical artist, the conclusion could be drawn that structured worship practice ought to be a holistic combination of scriptural reliance within a modern cultural context.

FILLED WITH THE INDWELLING
OF THE SPIRIT OF CHRIST

Each verse begins by calling on one person of the trinitarian Godhead:[83] Colossians offers an imperative action to "let the word of Christ dwell richly in you," and Ephesians the passive "be filled with the Spirit." Each suggests it is something that believers cannot accomplish themselves. The phrases are in the present tense, denoting an ongoing, contemporaneous action, not a one-time prior action. They are plural in nature,[84] offered to the corporate church as a worshiping community and not the individual believer. Each phrase presents a dual role assigned to a risen and active Christ, still working in and through the church body. In this way, these two phrases are related thoughts with regard to the effect upon the corporate body: the "word" is the "message" of Christ as taught and admonished to one another, while the "Spirit" is God overtaking a believer upon baptism upon conversion.[85] The effect of teaching and admonishing through psalms, hymns, and spiritual songs is the message of Christ—who he is and what he has done—indwelling and filling with the Spirit in the hearts of the congregation.

LET THE WORD OF CHRIST DWELL RICHLY

Throughout the Bible, the phrase, "word of Christ" is only found here

---

[83] Through there is a triune sense with God, Christ, and Spirit present, Paul is concerned with Jesus. Believers are to focus on the "word of Christ" and have their actions performed "to the Lord."

[84] Even though they may still apply to the individual, in the text these are given to the corporate body of the church, not to the individual Christian. It is a common error in the scholarship today to read "in you" and presuppose it to be a command to the individual, when it is a plural "you."

[85] Crabtree, "Colossians 3:16," 16.

and in Hebrews 6:1; the more common phrasing is "Word of God."[86] There are a few competing schools of thought regarding the meaning of "word of Christ": (1) the Holy Scripture;[87] (2) the actual spoken words of Christ; (3) the teachings or message of Jesus; or (4) a combination of these. Generally, scholars believe the overall idea of the letter favors the objective genitive use over the subjective genitive, and therefore the "word of Christ" might better fit the idea of "message of/about Christ." Because the text is open to all possible options,[88] by concluding the verse with "to God," it is both subjective and objective. In this way, Paul is specifically emphasizing that which comes from Christ himself is also to be used for God and building up his people. This fits with the church's need to understand the effect that the gospel is to have on the regular daily actions in their lives adding to Paul's call for the congregation to hold a high Christology. There is not much support for the meaning being Christ's exact words and better fits the context if it refers to the truths about Jesus, referring to the gospel in its broadest sense, encompassing all doctrine produced from the person of Christ. In fact, Colossians mentions Christ rather than Spirit because the focus of the epistle is his centrality and supremacy. Realities regarding Christ's being and actions are easier to share within "spiritual songs" as charismatic outpourings of the Holy Spirit.

"Dwelling" is a predominant theme throughout Scripture. In the OT, the Israelites sought to dwell in the promised land as well as have YHWH dwell with them in the tabernacle. In the NT, the believer permanently receives the indwelling of the Holy Spirit of God upon becoming a believer. "Dwell" is best understood as "in their midst"—similar to Hebrews—which fits the corporate idea of the message overtaking the whole of the congregation. "Notice that this dwelling is not automatic. Believers must co-operate in the Christian life as they do in salvation."[89] There is a bi-directional role in this dwelling. Where the indwelling of the Holy Spirit is an act of God upon faith in Christ, the dwelling of the word of Christ comes when the congregation is teaching

---

[86] Utley (*Paul Bound*, 45)notes: "There is another Greek manuscript variation here which is like that in v.13 and v.15. Scribes tended to unify Paul's expressions. The phrase 'word of Christ,' is a unique expression found only here in the NT. Therefore, it was changed to 'word of God' (MSS A, C*) or 'word of the Lord' (MS ℵ*). By far the best Greek manuscripts, P⁴⁶, ℵ², B, C², D, F, G, and most ancient translations have 'word of Christ.'"

[87] This belief is rare but held by some. If the understanding of "word" is meant to be *logos* in its strictest sense, it could be confined to Scripture alone. This is problematic when seeing the text from the point of view of Col 3:16, as it would limit the singing to only the Psalter and leaves no room for hymns or spiritual songs, even if it could fit the totality of twenty-first century canon.

[88] Moo (*Letters to the Colossians*, 285–86) suggests Christ is the object of the word: "Probably Paul means not 'the word, or message, that Christ proclaimed' but 'the message that proclaims Christ.'" Paul is referring to the significance of an authentic teaching about Christ. Likewise, Moo notes that *Christos* in the HCSB is translated as "Messiah," which is an important distinction when seeing Paul's eschatological point of who Christ is, the anointed and expected Jewish King and Savior. Viewing it in this way further works to tie the gentile church to its Jewish roots.

[89] Utley, *Paul Bound*, 45.

and admonishing through performance ritual. Paul intends a permanency, even though there is still action to be performed by the congregation. The Christian who truly follows Jesus will consequently teach, admonish, sing, and worship, thus ensuring the congregation is continually filled with Christ's message. Beyond simply dwelling, it is to be dwelling "richly." This suggests a deep and penetrating indwelling that brings a joy from making the word the focus of their lives.

Paul's citation of "word of God" is the important defining factor on how to understand the phrase "psalms, hymns, and spiritual songs," because if believers understand it to be the message of Christ, then the songs cannot be limited to simply the Psalter or just traditional hymns. All worship that preaches Christ would be acceptable for allowing the word to dwell, with the present imperative denoting an ongoing obedient nature. Letting the word "dwell richly" means to let it grow within believers daily, perverting neither their worship practice nor Scripture itself for one's own purposes or the purposes of the church. Paul appears to demand that the congregation put the message of Christ at the center of all their corporate worship, including within contemporary technical arts. Where producing worship is, so too should be the message and gospel of Christ.

BE FILLED WITH THE SPIRIT OF GOD

The phrase "be filled" (*plērao*) in the book of Ephesians is not the same "be filled" (*pimplēmi*) as in the book of Acts. The result is different. In Ephesians it is a person-oriented governing of one's life, while in Acts it is a task-oriented outward evangelism through tongues, prophecies, miracles, healings, and signs and wonders. Though it could be said that the response of being filled is speaking, singing, making melody, giving thanks, and submitting because those all modify the verb "be filled," that misconstrues the full understanding. The verb is a passive imperative, meaning the congregation must continually do something for something else to happen to them. Being filled with the spirit of God happens after one performs the actions; it does not create them. Likewise, it is a command, not a suggestion. There is an interesting theological paradox here. The filling of the Spirit throughout the NT is a one-time filling. Those following the command would already be filled with the Spirit in being believers. In this way, Paul must be presenting a different action of the Spirit in a congregational sense. He is offering a metaphor for what it looks like when the congregation is in unity, performing outward acts of the Spirit already held internally.

The surrounding verses in Ephesians 5:15–20 contrasts wise living with foolish living. The fool is moved by drunkenness and the wise by the Holy

Spirit. While filthy drinking songs were common in the Hellenistic world—to which drunkenness would lead—the Spirit leads believers to perform psalms, hymns, and spiritual songs. "If the Spirit is the source of their fullness, then, instead of songs which celebrate the joys of Bacchus, their mouths will be filled with words which build up the lives of others and bring glory to the living and true God."[90] Whereas being drunk is neither good for the person nor those around them, the Spirit is good for both. The distinction is not so much a contrast as it is a similarity: both wine and the Spirit are controlling factors in a believer's life. A person is "under the influence" of one or the other. It is the result that changes. The Spirit builds up; drunkenness tears down. There is another distinction between the two types of filling: at Pentecost the crowd thought the disciples were drunk on wine rather than understanding they were being moved by the indwelling of the Holy Spirit. In this way, the control of the Holy Spirit on the believer commands proper worship actions so to not misconstrue intent to the non-believer. The impact is filling the corporate body with the Spirit. The filling is not in a charismatic sense—though it could include that—but rather an ongoing presence of the Holy Spirit mediating a Christ-changed life.

One note here about a popular practice in creating an environment for worship in regards to the Holy Spirit. Among techs, there has been noted a semi-popular practice of lowering the air conditioning to create the "feeling" of the Holy Spirit working. The technique is to "blast" the chills as the band would reach a popular bridge section. The effect is a congregation who feels "moved" by the Holy Spirit. Even though environmental manipulation is an important aspect of the technical arts, the texts suggests the use of such methods should not be a substitute for the actual working of the Spirit within the congregation. "Being filled" is a continual gearing of one's life toward the person of Christ and not a moment of experiential "chills," though it could include that.

SPIRIT AND CHRIST CONCLUSION

Throughout the Pauline literature, dwelling always relates to the work of the Spirit. When the Spirit is active within the church, Christ's message resides in their midst. Even though the terminology is different in "dwell" and "filled" between the two verses, they are two sides to the same coin. Focus on the gospel message results in being filled with a dwelling of the Spirit and a heart that praises the Lord through building up the people. Colossians and Ephesians both connect the ministry of the Holy Spirit and the function of

---

[90] Bruce, *Epistles to the Colossians*, 380.

the "word of Christ"[91] with Christ's message dwelling within the midst of the congregation wherein a lifestyle change takes place once the believer is filled with the Spirit and lets the word dwell richly. The result is teaching and admonishing through the performance of psalms, hymns, and spiritual songs.

In practice, the passage suggests that it is not only pastoral leadership who have an active role in allowing the word of God to dwell and who assist in the Spirit to reside within the corporate body. Technical artists, likewise, perform a key role during the congregational worship experience. In practice, technical artists directly influence how the congregation will grasp the message presented. Where the pastoral leadership may define the church's vision and be ones who physically speak from the stage, technical artists interpret the message in terms of an auditory and visual sensory presentation so that the meaning that is finally received by the congregation is a combination of both the pastor and the technical artist. In this way, teaching and admonishing through psalms, hymns, and spiritual songs demand a response by the church technical artist to understand how their actions affect the intended message and what is being delivered to and experienced by the congregation. Just as they can assist in allowing the word to dwell and Spirit fill, they too can be obstacles against it when not using their craft as a supplement. All artistry will instruct; the question is, what is it teaching?

WORSHIPING COMMUNITY

The purpose of the letter itself is to meet an ecclesial need. The use of psalms, hymns, and spiritual songs is to communicate the beliefs of the Christian religion. Understanding the passage as communal means that those in charge of worship are responsible for teaching and admonishing along with the pastoral leadership. Because the purpose of an AV system is to share information and meet a communication need, it must fit within the environment and not be the focal point of it. Paul seeks to offer direction corporately for the church and to steer them toward Christ-like living, not just a turning away from their false beliefs. Thus, Paul uses produced worship as the means to establish the community's identity. Having a collective, structured worship practice builds up the body of the church.

---

[91] Risbridger ("Ephesians 5:18–21," 242) suggests that it is important not to presume the word is Spirit and the Spirit is the word, even though Scripture has similar parallels. They do work in close relationship with each other and lead to a dynamic working of God within the person and congregation, but in this instance "word" does not represent the "Word" in the same way John referred to it in his Gospel. Here word is not God himself, but his teaching.

EDIFICATION THROUGH TEACHING AND ADMONISHING

Charles Spurgeon wrote that he used to believe that no song should be sung unless it pointed toward God. However, Colossians 3:16 changed his thinking because of the verse's suggestion that songs can also be used to teach and admonish for the good of the members of the church. Charles Spurgeon suggested that while preaching it can be helpful to use verses from hymns. Rhyme and rhythm make it easy for people to remember; the congregation is able to understand the message better when they leave with a quick melodic line of truth. He suggests that sometimes the best way to admonish someone without their noticing is to sing a hymn with the correct truth attached. A preacher can teach "accidentally on purpose."[92] Jeff Crabtree concurs: "The instructions demonstrate biblical support for the belief that the Church should sing congregationally and that singing should be done with at least two purposes: edification and worship."[93] Instruction and building up the church is not only a role for the apostles and church leaders but the entire congregation as well. Teaching and admonishing comes not from church leadership down, but horizontal among them for corporate mutual edification. The performance of psalms, hymns, and spiritual songs is not intended to merely stir the Spirit but rather to form a biblical imperative that involves edification.

Both "teach" (*didaskō*) and "admonish" (*noutheteō*) are found earlier in Colossians 1:28 to describe Paul's ministry. The parity suggests that that which is good for Paul's ministry is also good for achieving the church's and congregation's greater purpose among one another. Richard Melick writes, "Teaching is the orderly arrangement of truth and effective communication of it. . . . Admonishing has the element of strong encouragement. It is generally practical and moral, rather than abstract or theological."[94] Teaching is the positive presentation of truth, while admonishing is the negative warning of straying away from the message of Christ; both are to be done in wisdom. Musically, teaching and admonishing cannot be separate from the act of worship; they must be inclusive means of the congregation's response to God and one another.

Songs in the ANE were transferred via personal interaction and not necessarily in written form.[95] Each person learned from one another as the medium for transferring truth through both words and cadence. Likewise, prior to the printing press, music and hymnody were common methods for passing along instruction, and served as the customary method in Greco-Roman

---

[92] Spurgeon, *Commentary on Colossians*, 14662–89.

[93] Crabtree, "Colossians 3:16," 18.

[94] Melick, *Philippians, Colossians, Philemon*, 304.

[95] This explains the lack of written evidence regarding song structure and notation beyond stanza markers and general stylistic direction.

society. Thus, the church does not learn only through preaching but also through both the action and content of song. It is a participatory activity, internally instructing the song's message from the heart and externally sharing through voice. Performance worship is not intended to be an emotional release, but rather a method of biblical instruction both for the worshiper and the congregation among them.[96]

Recent studies in NT hymnody show that both didactic and hortatory elements were featured, and thus teaching and admonishing were known attributes of hymnal singing. Richard Melick writes:

> Paul did not identify music as a spiritual gift, but he omitted other talents as well. This passage teaches that the spiritual gift is not music, but music may become an effective vehicle for the exercise of a gift. The gifts are teaching and admonishing. The medium of music, therefore, must remain secondary to the message it conveys.[97]

Teaching through art and song was a common practice, and Colossians 3:16 and Ephesians 5:19 offer the biblical foundation for it. In modern society, nothing is a more effective vehicle for teaching and admonishing than the living example of those singing demonstrating the grace in their lives that they feel from a changed life in Christ, moving beyond itself, exhorting and encouraging believers and non-believers to focus on Christ. Performance arts are themselves the spiritual gifts because they serve as the vehicle that makes the ability to teach and admonish effective. In this way, church technical artists can teach and admonish when they use their art to demonstrate the grace in their lives as they lead the congregation to focus on Christ through the act of worship.

BI-DIRECTIONALITY OF
PSALMS, HYMNS, AND SPIRITUAL SONGS

Psalms, hymns, and spiritual songs are a mutually edifying vehicle for instructing one another and in praising God. While Pliny understood early Christian worship as bi-directional—both antiphonal to one another and to the Lord—even today our hymnbooks are dual focused, with vocal and musical representation collectively vertical to God and horizontal among the congregation. Thus, there are two audiences, God and "one another." The purpose is to build up the community. The reflexive pronoun in v.19 for "to

---

[96] Aniol (*Sound Worship*, 10) suggests the reference to the word of God dwelling is not talking simply about teaching truth of God but says something about the power of music, namely that it itself has the power to teach. The content itself dwells within the individual through the act of worship. He points to the Jewish hymnal as an example of songs that can themselves fill the volumes of a systematic theology.
[97] Melick, *Philippians, Colossians, Philemon*, 305.

each other" demonstrates the object duality. In the course of worship, Christians are speaking to God and each other. When a member of the congregation is able to see and hear the congregation worship, he or she is also drawn into the experience alongside them. Technical artists create the experience for the message to become the physical reality of the sung truth.

In regards to early church worship, it is thought that "to one another" hints at antiphonal singing, with one person singing one hymn, then another person either repeating it or performing the next line. Then, back to the original singer, and so on. The meetings are then as much for social enjoyment as for the public worship of God, exactly because antiphonal singing engages the audience through active participation.[98] It is a joining of shared theology through voice and music. A parallel is to that of national anthems we sing today. It is natural for people to join together in a communal song in order to feel the unity of patriotism. In this same way, psalms, hymns, and spiritual songs connect the body of Christ in a common purpose of praise. Singing to one another suggests the need for congregational gathering and support of the church.

> Christians filled with the Spirit may play an educative role in training each other in the proper praise of God. Thus, in praising God in a public communal way, Christians are also indirectly addressing each other, encouraging and instructing each other in what will be their eternal practice, and thus building up the church.[99]

The singing of praise in the Bible is always done to God, but this passage adds a horizontal dimension in that when the congregation gathers, their individual praising God encourages one another through a corporate focus on God. The vertically focused action generates a horizontal affect.

The fact that there was singing in the early church is not astounding, but what is surprising is that in doing so, social boundaries were broken. In worship, unified action is created among Jew, gentile, aristocrats, slaves, free people, male, female, young, and old. They all come together in one voice for Christ. It is a harmony of brothers and sisters, formerly enemies, now brought together in and through Christ. When among one another there becomes a responsibility for one another. The message in song cannot be frivolous or light in the message of Christ because everyone is responsible for everyone else's growth toward Christ. Submitting to one another is the focus of the following pericope, though it is connected to the former imperative. Therefore, singing and submitting are not exclusive of one another. They are joined. It is a submission that creates synchronicity. Worshipers submit themselves to a common structure, both musically and stylistically.

With Colossians 3:16 focused on the church—with *en* used as "in" also

---

[98] Spence-Jones, *Colossians*, 210–11.
[99] Fowl, *Ephesians*, 178,

translated as "among"—Paul emphasizes its congregational importance. The unifying agent is the Holy Spirit. It is the source of praise directed vertically "to God." Worship, therefore, is two-dimensional due to the work of the Spirit within the changed heart of the congregation. Through communal worship, the congregation reminds one another of the actions of God, thus adding to the significance of the experience. In Paul's writing the Holy Spirit speaks out words and inspiration that point to Christ and submitting to his authority. It homogenizes and opens up the body of Christ for love of one another. In this way, just as God is unity as three-in-one, believers too become unified with one another and God through Christ. Jeremy Begbie summarizes well:

> "The pattern is to God (the Father), in the name of as well as through the Lord Jesus," while in Ephesians, it is "a similar pattern ('to God the Father . . . in the name of our Lord Jesus Christ') . . . with the addition of singing to the [exalted] Lord."[100]

In this way, it can be suggested that the early church could have likely believed that offering praise to God through Christ held a higher purpose in servicing the spiritual needs of the church itself.

In a practical sense, InfoComm International's *AV Solutions Provider Standards of Excellence* notes that all aspects of creating an AV experience—from scope-planning, training, integration, and business practices—are to holistically meet the communication needs of the client. For church technical artists, the communication need is not solely technological but bi-directional, connecting congregant-to-God and congregant-to-congregant. All practices are to be aimed at that end. The text thus suggests that technical artistry that is aimed at God through presenting the word and truths about Christ meet that standard by building up the church. The "one another" who is the technical artist builds up the "one another" who is the congregant.

A HEART OF WORSHIP

The phrase "in your hearts" denotes a sincerity and devotion to worship, expressing how the performance of psalms, hymns, and spiritual songs is to be expressed. The vertical aspect of worship begins with the self in response to "putting on" a heart for Christ. An internal attitude creates an outward expression demonstrated through structured liturgical performance. The internal desire to sing out is the true-tell sign that one has Christ's love within his or her heart: it is not a silent, internal singing but rather singing that comes from the depths of where Christ resides. The heart is the focal point of music;

---

[100] Begbie, "Temples of the Spirit," 71.

it reflects the believer's affections toward God. Corporate worship brings together both inner and outer lives, being able to connect one another in a way that is not possible alone. "Singing enables the expression of the individual heart, the encouragement of fellow believers, and, overall, the praise of God."[101]

The edifying role Paul is speaking of correlates well with the Greek understanding of the power of music. Music could change behavior, influence morality, and was concerned with truth and beauty. Music was much more than just inspirational praise; it served as a means of instruction for one another as well as mass edification. The heart itself is the actual "instrument" or "region" from which song derives. Being based in the heart does not mean that it is only an emotional response, though it is that. The heart is where the inner self of a person lies which in turn impacts the intellect and ability to reason. The experience is not a charismatic, emotional one—though it can include that—but heart-led act of reverence in wisdom. It allows for a giving of the whole self and not simply a vocal outpour. Even if spontaneous, it is a reaction to a changed heart of thanksgiving. It is the internal motivation that provokes the external response. "The heart goes before the tongue," wrote John Calvin.[102] As a contemporary example, Reggie Kidd draws the picture of "Bubba," an average guy in the congregation who does not bring anything more to the service of his church than a clean heart desiring to praise and proclaim Christ. Still Bubba is able to lead and teach others effectively through his craft.[103] Christ's work perfects the worship acts of the average, the rich, the poor, the downtrodden, and the Bubbas because when Christ built his church from the ordinary, from the masses, hearts change. When church sound booths are filled with heart-led volunteer teams, Christ is preached through the audio, video, and lighting experience.

WISDOM

Wisdom is not commonly thought of as a natural component to singing in everyday life, yet Paul claims it is essential elements to teaching and admonishing through psalms, hymns, and spiritual songs. Wisdom and understanding are evident through sharing a deeper awareness of the community's practices. It is a "participatory knowledge":[104] the believer learns what it feels like to experience God.

Wisdom is a common theme throughout the Pauline literature. In both

---

[101] Stearns, "Participation," 3653–54.
[102] Calvin, *Commentary on Colossians*, 3751–53.
[103] Kidd, *With One Voice*, 147–60.
[104] Guthrie, "Wisdom of Song," 395–96.

Colossians and Ephesians Paul presents psalms, hymns, and spiritual songs as methods for gaining the wisdom of God while practically demonstrating that gained wisdom through the acts of teaching and admonishing the congregation Wisdom is always presented as possessing a spiritual dimension related to the mind of God. Its use is a literary play against the false wisdom presumed by the Colossian Heresy. For Paul, true wisdom comes from Christ, not man. That wisdom is "intellectual" because teaching and admonishing both come from the internal knowledge of the message of Christ that a believer possesses. "The fact that Paul ascribes a strong teaching function to the lyrics of early Christian hymnody . . . stresses that there was also a strong cognitive dimension to the singing in worship," notes Clinton Arnold.[105] Paul appears to contend that worship is an intellectual, not emotional, endeavor as the means to deliver the message of Christ to the congregation. Though the effect on the person can be emotional, the wisdom shared is centered around scriptural truth that fosters from a converted heart indwelled by the Spirit.

Practically, for the artisan creating a technological environment, wisdom becomes the defining factor: tech managers must consider the ultimate effect before any action takes place. They must consider the usefulness, usability, and adaptability. Technology should be purposeful and well thought out. It is to be intentional even while being artistic. A popular tactic is the creation of the "bumper video," a mini-movie of two to five minutes that plays between the worship set and the pastoral message. It serves to set the mood for the upcoming sermon while calming down the emotional outpouring from the worship experience. Wisdom would appear to dictate that the videos not be haphazard nor purely entertainment nor emotional, but rather bridge the spiritual truths just sung with the theological foundation of the pastoral teaching to follow. Wisdom dictates that the mood created by the background music, lighting, and production style move the congregation to think theologically.

SINGING WITH THANKFULNESS

Thankfulness is a secondary theme carried throughout the book of Colossians. Though the word is often translated as "with gratitude," which is lexically possible and theologically consistent, this is never done elsewhere in the NT. David Detwiler suggests that better understanding is "God's grace."[106] It is not so much about believer's graciousness toward God, but the grace he showed them. It is about God's actions, not the worshiper's.

---

[105] Arnold, *Ephesians*, 361.
[106] Detwiler, "Church Music and Colossians 3:16," 363–64.

Though Detwiler's suggestion is theologically valid, it does not consider the complete the context of the passage. Here, it is about both God's actions and the congregation's, expressing how they are to respond to God in worship, namely with thankfulness in their hearts. The passage concludes with the imperative that the congregation must do everything with thanksgiving that is pleasing, profitable, and in the name of the Lord Christ. In the NT, there are no lists of dos and don'ts, but actions performed due to a heart of thanksgiving; one puts-off old behaviors and puts-on new ones found in the identity of Christ. While throughout the Psalter giving thanks is a common theme, here thanksgiving is not likely a reference to any one particular practice within their worship service, but a stance of gratitude.

Similarly, social giving was commonplace in the Mediterranean antiquity. The material act of giving was the method for demonstrating thankfulness to someone or for something. When given a gift, the proper response was a reciprocal gift. In the case of a believing community, the gift of salvation and redemption leads to a giving back in thankful praise through psalms, hymns, and spiritual songs. Outward thankfulness is to be demonstrated visually and verbally through a worshiping congregation.

The text suggests that to Paul the root location of thankful song is the heart, which in practice is vocalized and visualized.[107] Spirit-filled thanksgiving generates an expressive, spontaneous response in song and worship. When working through the believer, the Spirit creates a thankful heart, shown through praising God. The word "singing" that opens the second half the verse carries instrumental force, showing it as a means by which something is to occur. It shows that thanksgiving in the heart is expressed both internally and externally. Outwardly vocalizing thanks to Christ demonstrates that the congregation is not ashamed and serves as a declaration of Christ as central in their lives. In this way, thanksgiving cannot be offered, Paul exhorts, without also edifying the congregation.

WARNING: HEEDING TO THE HERESY: SERVING GOD OR ENTERTAINING CULTURE?

Even though the exact heresy, or heresies, are unknown, based on the context it no doubt included influences that encouraged conforming to the pagan culture. When one puts on worship practices that teach and admonish, he or she puts off the desires of their societal ethos. Yet, the old and new self are in constant battle. For the technical artist who works in the realm of embracing cultural mediums and transforming them into edification of the

---

[107] There are multiple biblical parallels about that which comes from the mouth is an expression of the heart: Luke 6:45; Matt 12:34; 15:18; Prov 4:23; 13:3; 18:21; Jas 3:5b–8; and Eph 4:29.

church body, there is a continuous struggle between being socially relevant, trendy entertainment, or true to a changed person in Christ. Kevin Vanhoozer suggests, that church artists do "is always educating; the only question is, 'What is it teaching?' . . . What norms, values, and belief is it conveying through its hidden curriculum, its everyday ways of doing things?"[108] Therefore, technical artists need not only be aware of the way they perform their practice but how their product will in turn influence and instruct the congregation. Church tech does not simply manufacture a product, it transforms subjects.

Viewed in this way, technological productions can become a substitutionary "Bible for the post-literate."[109] Due to the current generation's trend of exchanging paper Bibles for iPad apps, laptops, Bible software, social media, on-demand instruction, and online video campuses, what they hear through speakers and view on screens is often the only way they will learn the Bible's teachings. As the culture becomes more accustomed to computer-mediated communication, technical artists themselves become the administrators of the effectiveness of the communication device as well as the message perceived. For this reason, some argue against the practice of transitioning away from printed works and toward projection screens, suggesting that the written word is a more effective method for contemplating truths of the Christian faith due to the additional accountability in content selection by having to go through a professional editing process. Likewise, on-demand media can be provided in response haphazardly in order to provoke feelings of cultural acceptance rather than cautiously presenting biblical truth. David de Bruyn contends that projected lyrics do not offer reflexive theological context and come dangerously close to falling victim to a "whimsical" on-the-fly interpretation of the laptop holder. [110] In this way, the internal motivations of the technical artist have a distinct impact on the presented song, and thus carry an added responsibility. A level of responsibility not all are ready to give.

Though Paul is focused on putting off sinful habits (3:5–8)[111] and finding knowledge from God above (3:10), the use of old and new self is less about exact virtues and vices and instead express a complete change in identity from being made in the old Adam to new in Christ. This is important because it expresses status: the believer is the new self and therefore his or her actions should be representative of that distinction. Schultze, Chuang, and Redman note how those in the church view technology; they "tend to use language that describes technology as neutral. The common metaphor used to talk

[108] Vanhoozer, *Faith Speaking Understanding*, 147–49.

[109] Schultze, *High-Tech Worship?*, 19.

[110] David de Bruyn, "Thanks, But I'll Keep My Printed Hymnal," *Religious Affections Ministries*. February 9, 2012. http://religiousaffections.org/articles/hymnody/thanks-but-ill-keep-my-printed-hymnal/.

[111] Of sexual immorality, impurity, passion, evil desire, covetousness, idolatry, anger, wrath, malice, slander, obscene talk, and lying.

about it is a tool: it has no moral value in and of itself but is entirely about how it is used."[112] They continue; it is much more than a tool. With the power to express both sides of their nature—before and after Christ—the technology will take on the character of the one who utilizes it and can still present the message from the background of which it was derived, even if used in a different context. Bob Kauflin suggests that bringing the secular into the religious risks failing to present the gospel message the congregation comes to hear. Even though the reasoning for relating to "seeker" attendees is to make them feel comfortable, the purpose should not be comfort but changing hearts; incorporating everyday music and technology risks making people who are seeking to find help leave due to the church appearing like the same places they feel hurt during their regular life.[113] Kesha Williams's and Omotayo Banjo's study of Contemporary Christian Music listeners discovered this same conclusion. They found the primary motivation for listening to CCM was "to connect with God . . . as a substitute to unproductive messages associated with secular music and it reinforces listeners' beliefs. In so doing, listeners are able to maintain their faith and . . . relationship with God."[114] They discovered congregations listen and participate in Christian music ritual in order to remain focused on Godly living throughout the week, and most notably when secular aspects of life like work and pop culture steer them away from the desire to live holy. In this way, William Willimon warns that rather than calling the conforming to modern worship practices apostasy, "all too many of us label it 'contemporary' and 'relevant,' . . . [and] no one calls it faithful."[115]

Likewise, technical artists themselves risk being influenced by the technology and society from which it derives. One finding in the documentary analysis was that attendance in secular music industry trade shows outweighed attendance at church tech conferences, often with the rationale being that that is where the new tools of the trade are brought and therefore where attendance is required. In fact, not one church tech article presented an argument not to attend or incorporate secular events or technological concepts; every article that referenced secular trade shows gave justification for using them. Some would claim they are being harnessed for God's purposes or as an exploration of the world God created, but none provided warnings against the possibility of negative influence of the secular on the artist or congregation. However, the research found that the same vendors and manufacturers attend both secular and Christian events, which makes the

---

[112] Schultze, Quentin J., D. J. Chuang, and Robb Redman, "Worship, Technology, and the Church: A Discussion with Quentin Schultze and DJ Chuang," *Cultural Encounters* 1 (2012): 103.

[113] Bob Kauflin, "Should We Use Secular Songs on Sundays?" *WorshipMatters.com*, August 17, 2007, https://worshipmatters.com/2007/08/17/should-we-use-secular-songs-on-sundays/.

[114] Kesha Williams and Omotayo Banjo, "From Where We Stand," 206, 209.

[115] Willimon, *The Bible*, 15.

reasoning for attendance incompatible with the reality. In this way, "making good decisions about our participation in modern life requires having a clear understanding of our relationship to technology in the context of our relationship to God."[116] Ron Man refers to this as "whole-life worship,"[117] in that being consistent with the use of shared artistic mediums requires a disciplined and intentional separation not found in the documentary analysis of technical artists.

Church tech director Stephen Beasley admits: "As a younger media director in a big church I have known the adrenaline of a great show. I've pushed a talented team to limits only NASA could rival. I have witnessed pure audio, video and lighting ecstasy spring forth from buttons and faders."[118] (2015, no pagination). Beasley concludes that it made him feel empty and that the practice strayed from pursuing Jesus. The Colossians text suggests the reason is that these incorporate activities aligned with conforming to the heresy and practices noted in the put off section, namely idolatry of aligning with what is deemed culturally acceptable. In the same way, Paul exhorts the congregation to put off anger, wrath, malice, and obscene talk. Yet, church techs are often known for being the most negative people in the church: they are the department of "no," often destroying relationships rather than embracing them. Technical artists often believe they are able to hide in the back, lay low, and not have direct contact with the congregation. If this is the attitude embraced while serving in their capacity, and they are the ones using their medium in order to teach and admonish, then it could be argued that the negative habits and cultural stigma attached to the secular practices brought into the worship service are the same traits being taught to the congregation. Therefore, maintaining a high value on teaching and admonishing the message of Christ to the congregation through a changed heart for Christ is to remain central to every creative activity.

CONCLUSION

Teaching and admonishing are the actions. The method is singing and performing psalms, hymns, and spiritual songs. This is done to God with a heart of gratitude and thanksgiving. Together, eleven guidelines can be exposited from the parallel Colossians and Ephesians passages which all work together to create a holistic worship experience for the NT church:

(1) The plurality of psalms, hymns, and spiritual songs suggests that the

---

[116] Ermer, "Responsible Engineering and Technology," 132.
[117] Man, "Biblical Principles of Worship," no pagination.
[118] Stephen Beasley, "Backtalk: People vs. the Product," *Church Production Magazine*, Accessed June 7, 2015, https://www.churchproduction.com/ministry/people-vs.-the-product/.

medium is not just "singing," but all artistic forms of worship aimed at praising God and building up one another that includes a purposeful, produced worship experience;

(2) Worship services are to be planned liturgically, including but not limited to psalms, hymns, and spiritual songs, with each element teaching something of Christ and right Christian living;

(3) The music and production are to be artistic; not about style, but substance;

(4) The experience should possess an element of admonition and specific warning against false teaching and practices;

(5) The message must be understandable, intelligible, and thoughtful;

(6) It is to be edifying to one another, for one another, lifting up the corporate body from a heart filled with thankfulness and gratitude;

(7) Songs are to be spiritual, directed toward God, and Christocentric;

(8) The result must be a corporate filling of the message and word of Christ filling the minds and hearts of the congregation;

(9) Actions result from motivation from the Holy Spirit;

(10) When done properly the message of Christ and Spirit of God will be present among the people; and

(11) There is a warning against idolizing popular culture and the possible negative influences of integrating secular practices.

In this way, the technical arts ministry becomes a bi-directional facilitation between God and the congregation through the word dwelling and Spirit filling and serves these purposes when presenting the message of Christ through production arts. Technically artistic worship would then be a possible method for the church to let Christ be known both internally to one another and externally evangelistically. It not only kindles the hearts of the congregation but helps the church grow in wisdom and understanding of God. It promotes corporate worship that is both of the heart and of the mind. "Psalms, hymns, and spiritual songs" demands that the method for accomplishing this is through a multi-modal and multi-medium worship. These findings, therefore, suggest that Christian worship would be well served when it encompasses Scripture, historical church hymns, the heart of the church artisan, and performed bi-directionally—to God and to one another—through the creations of the contemporary church technical artist. Technically based psalms, hymns, and spiritual songs are the vehicle for teaching and admonishing the modern church; the effect is the word of Christ dwelling among the body of Christ. It is a ritual practice aimed at turning attention off oneself and worldly ways and onto Christ for the edification of the congregation.

❧ 6 ❧

CONCLUSION: "PRODUCING WORSHIP"
A THEOLOGY OF CHURCH TECHNICAL ARTS

A TECHNICAL ARTS METANARRATIVE

The historical progression of worship and creative arts practices aligned with the biblical principles explored in the passages from Exodus, Hebrews, and Colossians can be read together to paint a vivid picture of the people charged with the production of church worship services and the qualities they are to possess. This assertion is significant specifically because the passages themselves are not traditionally understood as worship focused, nor would historical traditions have been able to hold technological advances in view. The theology constructed from these passages offers a holistic approach to modern church technical arts wherein Spirit-led artisans skilled in knowledge and craft (Exod 35:30–36:1) facilitate the worship between God and humanity through the band and stage as Christ's mediators (Heb 2:12–13) for the purpose of building up the corporate body through teaching and admonishing one another, via multi-sensory, artistic mediums (Col 3:16). In this way, church technical artists produce worship for the church that mirrors their own worship as an exemplification of themselves as image bearers of Creator God. An inward stirring of the heart creates an external outpouring of sound, light, video, and atmosphere that shapes how the congregation experiences their sacred space.

Historically, art and music either directly mimicked or diametrically opposed societal norms. The struggle was real. However, in all instances worship practice would eventually take on the identity of the musical and artistic practices of the people. Even the Iconoclasts could not suppress the free expression of the Reformation and Renaissance. Whether it be the OT era

171

use of vocal praise, instrumentation, celebration, lament, and thanksgiving as presented throughout the Psalter and historical accounts or the contemporary clashing of styles within the traditional church and modern CCM and Jesus Movements, worship always took on the form of the people. Yet, common to all styles is the centrality of God. Where "authenticity" in worship production has come to be viewed as historical, a better understanding would be cultural. Gospel artisans throughout history have been able to harness current mediums in order to present God's theodrama in the most meaningful way for the congregation at hand. History holding true, contemporary church technical arts viewed both biblically and practically would be no different.

The tabernacle construction narrative is more than architectural plans. It is a blueprint for the character traits that those charged with creating worship within the church's worship space are to possess and utilize in practice. Those who are called by God fill a role of Bezalels and Oholiabs, producing the congregation's sacred space for corporate worship as commanded by God. Church technical artists whom God fills with the Spirit are to utilize wisdom, intelligence, skill, understanding, and craftsmanship while passing along that knowledge of both God and craft to the people of God. Vivid colors, sounds, and ritual filled the tabernacle in the same way modern technical artists fill the church sanctuary with a blending of notes and visuals. The text suggests that merely possessing the skill to operate the machinery is not sufficient. Excelling at one's craft requires gaining knowledge and understanding of all aspects, from the scientific principles of the mediums to the artistic interworking of the materials to understanding how each aspect affects the participant emotionally and physically, as well as how it relates historically to the faith. In the church—God's contemporary tabernacle—technical artists form the sacred space that allows God to finally dwell among his people, filling the place of worship with his glory, and drawing his people to himself.

In Hebrews, Old Testament faithfulness becomes the glory of New Testament worship as Christ himself leads the congregation from within their midst. God—as the Author and Perfecter of the faith—is the Singing Savior in whom David and Isaiah sought refuge. God's purpose was fulfilled when Christ humbled himself on the cross, placed lower than the angels but with the world as his footstool. Christ sings of his faithfulness to the Father alongside his brethren-children, leading them in praise as their High Priest. Just as Christ mediates the worship of the Father, so too do church technical artists via their physical participatory position mediate the revelation and response between God and congregation through the creative arts. In the same way that earthly pastors and worship leaders proclaim the name of God in the performance of their practice, modern technical artists lead the congregation's praise of God by facilitating the worship of the Father from stage to congregation and congregation to God as the message of Christ physically and digitally passes through the electronics of the tech booth from stage mic to loudspeaker. In doing so, technical artists serve a dual role on behalf of Christ, yet still, as one of the brethren-children he came to save.

In the book of Colossians, the message of Christ and filling of the Spirit become the defining characteristics of the NT church when the congregation teach and admonish one another through multi-medium worship. The singing and performing of psalms, hymns, and spiritual songs form a holistic approach to worship content and context that is found in the re-creation and representation of the biblical narrative through artistic mediums, by blending church and cultural history (psalms), new worship liturgy in praise and thankfulness (hymns), and spontaneous heart-filled outpourings controlled and directed by the Spirit (spiritual songs). In performing their craft, it can be said that technical artists present Christ-like, biblically focused qualities to the congregation, while at the same time, they put-off pre-salvific habits and put-on the characteristics of Christ. Their worship becomes bi-directional: technical artists serve their purpose when the congregation both increases in their knowledge of Christ and builds up one another in their daily lives. The Spirit continues his work when the word of Christ dwells richly among a worshiping congregation; the effect is a corporate body that grows in their faith and likeness of Christ.

Therefore, this constructive theology proclaims that, in practice, modern church technical artists serve as Spirit-moved mediators of bi-directional praise of God, through the person of Christ, as they produce aesthetically inspired liturgical applications of sound, light, message, and sacred space with wisdom, excellence, and craftsmanship. The Bible, then, can indeed be understood as saying something about "producing worship" and informing praxis for contemporary church technical artists. In this way, while technical artists serving in a similar role to artists of the past, they are also unique. Traditionally an artist creates a product that is itself an object for worship of God or that would be used to tell God's story. Today, technical artists utilize digital storytelling as temporal vehicles for the message of God; they are instrumental in the direct creation and presentation of the pastoral message. In no other artistic practices are the message and medium so connected. This suggests that the technical arts play a distinctive role in the church not seen since that of Bezalel and Oholiab.

Therefore, applying the exegetical portrait formed from each of the passages, certain conclusions regarding modern church practice can be drawn. Five common themes are woven among the three biblical passages: (1) the work of the Spirit; (2) mediation between God and humanity; (3) the creation of sacred space; (4) beauty and excellence; and (5) intentional workmanship.

THE WORK OF THE SPIRIT: LEADING ONE'S CRAFT
INTERNALLY TO PRODUCE THE EXTERNAL

The text suggests that the work of the Spirit is central to each of the narratives and can be viewed as essential to the performance of technological craft. In all three, the internal working of the Spirit generates the external actions of the creative arts. In Exodus, Bezalel and Oholiab are the first to be "filled with the Spirit" of God. The text not only states that detail but also specifically places the Spirit as the first of the creative artist's six required qualities. It can, therefore, be suggested that all the other qualities are to be viewed in relationship to the Spirit. Hebrews places Christ in the midst of the congregation, being the mediator between God and his people, which today is performed by a working of the Holy Spirit within the believer. Colossians suggests that "spiritual songs" are a specific way the body of Christ can be built up. These songs are to be both about God as well as focused toward God and stirred by the Spirit. As suggested by the parallel verse Ephesians 5:18–19, the effect is that the congregation is simultaneously filled with Spirit. When the believer's internal motivation is the responding to and focusing on the Spirit through worship, the church body is further united with God.

Biblical craftsmanship suggests cooperation with God and the Spirit. God filled Bezalel and Oholiab with the Spirit, stirred them up for the work, perfected their ability, intelligence, knowledge, and wisdom necessary to create, and placed them in a position to teach their assisting artisans both the craft as well as the truths of God. God endows both the Spirit and the artist's creative, technical skills in order to make a physical relationship with him possible. In this way, the gifts from God cannot be separated from the Spirit of God. The Spirit guides believers who are called to the particular work necessary to accomplish the task at hand. This presents a vision of technical artistry as both necessary and Sprit-led. God's active role in selecting the artisans as well as perfecting their abilities confirms the arts are a God-sanctioned activity when performed in accordance with the Spirit as provided by him.

> Though he [Bezalel] is commonly read as a prototype of the modern Christian artist, an accurate reading must understand his work not as an expression of creativity, but as a calling by the Holy Spirit to serve the believing community in their worship of God.[1]

Even though the text does say something about creativity—for the tabernacle is itself an extraordinary artistic feat—it further demonstrates that the tabernacle serves a more significant purpose of pointing the people to a redemptive future made possible through the Spirit's stirring.

Hebrews displays artistic creations as the present inspiration of Christ. Tangible and effective worship should not be performed as Christ-

---

[1] Harris, *The Curious Works of Bezalel*, 14.

remembered, but in and through Christ-revealed. God is revealed because Christ's work is present and ongoing, which today is through the work of the Holy Spirit within believers (John 14:16). Worship becomes the proper response for those under grace before him with the role of the technical artist serving as a practical way to demonstrate that response. The Holy Spirit within points church technical artists to the Father, transforming them into genuine worshipers of Christ. In this way, worship becomes faithful and appropriate. It can then be suggested that technical artists who are entrusted to the production of the worship service offer a vehicle for the Spirit to work among the congregation so that through Christ, the Father is praised and proclaimed. When technical artists see themselves as Christ's representatives—due to the indwelling of the Spirit—they demonstrate their trust in the relationship with Christ and the Father through obedience to his call. That confidence becomes the internal motivator, not shying away from trials but heeding the author of Hebrews's call to endure, having confidence that the Son through the indwelling of the Spirit is among them in the midst of the performance of their craft. Because Jesus shared in the human condition, he is able to assist them in enduring the trials of their craft, guiding imperfect human hands to create perfected worship that is acceptable to the Father.

Colossians 3:16 suggests that church technical artists, who are filled with the Spirit, are to teach the truths of the faith through their craft, musically and artistically performing psalms, hymns, and spiritual songs with thankfulness in their hearts. The work of the Spirit within the artist is the qualifying factor in producing the worship experience. The kind of music that Paul is referring to would be both vocal and communal, instrumental and multi-medium, and always aligned with the message of Christ and guided by the Holy Spirit. While the form is secondary to the theology, one specific type of worship is that of *odaes pneumaticaes* or "spiritual songs," which are spontaneous outcries generated by the Spirit within the believer. Worship becomes a matter of the heart, which has its end in mutual edification, not entertainment. In the put-on put-off section, Paul contrasts the positive actions of worship with the not-to-dos of fleshly and secular living. The focus here is a contrast between cultural norms and the ways of Christian life and religious practice. Adhering to societal norms is not the way to be filled with the Spirit, but rather by focusing on Christ through corporate worship. Paul is suggesting a liturgical worship practice wherein "do," and the Spirit fills; "do," and Christ dwells. The working of the Spirit is not passive but is instead a deliberate and embodied effort of the whole self, leading to building up the church.

The view presented here is that the Spirit ties together all essential activities necessary for worship. Knowledge of God and congregational edification become the products of worship perfected through Christ. When skill, intelligence, knowledge, and craftsmanship join together under the filling of the Spirit, the congregation responds by building up one another from the midst of one another through the acts of teaching and admonishing. The internal creates the external. When the technical artist responds to his or her call to serve, the role is to teach the characteristics of God to the church through their art form; the Spirit uses the skills of the technical artist from within the church and perfects the created technical production in a way that becomes useful and appropriate. The Spirit drives the technical artist to tangibly perfect his or her craft in order to present a worship service that is both a learned and spontaneous outpouring of the Spirit within.

MEDIATION: MINISTERING TECHNICALLY, TEACHING GOD TO HIS PEOPLE

Where the Spirit is the power of God working through the creative characteristics of both the technical artists as well as the hearts of the congregation, technical artists themselves have a specific role in administering the church's worship practice. They are the modern church's mediators between God and his people. Through the performance of their craft, technical artists present God to the people and his people to God. The sounds and visuals connect the stage to the congregation to prepare the people's response back to God: technical artistry is a bi-directional activity. Bezalel and Oholiab present the characteristics of God through their creation: through the building and re-building of the tabernacle, the Israelites are offered a mobile reminder of YHWH among them in recognition of the atonement made possible only through God's presence. In Hebrews, the author presents Christ as the Mediator of the people's worship of the Father from within their midst. Christ is among those who are given to him while also being one with his brethren. Christ serves a dual role of revealing God to his people and delivering the congregation's response to God as Apostle and High Priest. Technical artists mediate the sound, light, and visuals from stage to congregation within the midst of the people as one of those Christ came to save. The people of Colossae are to administer the knowledge of Christ among one another: one method of doing so is through teaching and admonishing by way of a multi-medium worship experience. The view here is that when psalms, hymns, and spiritual songs are performed corporately, the worship edifies the congregation and grows them in their knowledge of God. Church technical artists harness the music and message from the stage as the primary ingredients to teach and admonish, translating the presented words, sounds, and environment into artistic expressions that stir the people's hearts to outwardly live their internal transformation. Technical artists shape the way the congregation understands who God is and how they are to worship him.

In the OT tabernacle construction narrative, Bezalel and Oholiab are said to possess the ability to teach (Exod 35:34). The qualities of skill, intelligence, knowledge, and craftsmanship are the aspects by which productions ought to be judged, with the ability to teach as the judge of the technical artisan him or herself. That this detail is left out of the commissioning narrative in chapter 31—only added once the project is turned over to Bezalel and Oholiab—is significant. This demonstrates that when those overseeing the construction of the worship space are allowed to rely on God's gifting without interference, the impetus is to empower those who serve with them. When the project is in the hands of the leadership (Moses/church), the focus is often solely task and completion oriented, without understanding the greater narrative impact that the artistic qualities add. Through teaching their craft, technical artists fulfill their ministerial objective by passing on their worship practices to future generations to experience God, who then, out of those skills, can develop the necessary methods for engaging the changing cultural loci. Understanding that the tabernacle would be continually torn down, moved, and re-constructed, and that neither Bezalel nor Oholiab would eventually make it into the promised land, the passing down of the craft through a mentor-mentee relationship would have presumably been an essential part of carrying on the worship and liturgical traditions of the Israelites. In the OT, God chose mediators to speak for him, and even though the Levites would administer the word of God, it was the tabernacle artisans who would be responsible for providing all the necessary elements—from physical layout to the utensils, to the fabrics, to the ark, and all other physical characteristics—that collectively would represent God's earthly cosmos. Thus, those who provide the worship elements to the church today are also assisting in mediating the message of Christ to the people.

Throughout the NT, Christ is presented as the sole mediator between God and humanity in terms of salvation and reconciliation.[2] The book of Hebrews directly cites this as Jesus's role more than any other NT book.[3] The author connects former human mediators, like Abraham, Moses, and the high priest to the one eternal Mediator who sits at the right hand of God: Christ. Abraham and Moses were mediators from God's revelation to humanity; Aaron was the mediator of the people's worship to God. In Hebrews, Jesus becomes the once-and-for-all reconciliation between the Father and his people; both roles are filled by Christ. Whereas most NT citations relate Christ's intercession to the saving work done on the cross—which should not be overlooked—Hebrews 2:12–13 offers a unique example of Jesus's mediation in religious practice. Jesus takes the high priestly role from among the congregation rather than being separated from them. As High Priest, he pulls back the veil and brings his brethren-children with him into the true

---

[2] Cf. Gal 3:19–20; 1 Tim 2:5; Rom 8:34; 2 Cor 5:18–19; Col 1:20; 1 John 2:1–2.
[3] Heb 2:12–13; 4:15–6:1; 7:22–25; 8:6; 9:15; 10:19–22; and 12:24.

tent that is reconciliation with the Father. As the veil is torn, so too are the worship barriers: Christ's role amid the congregation forms the corporate worship bond, leading worship to the Father through song. With Christ's intercession, sinful man can boldly enter into God's presence, with Christ as the mediator of man's response to God. He does not only receive worship as God's co-equal, but he is himself the ultimate worshiper. Christ leads the people, making their heartfelt worship perfect. The text suggests that Jesus is not content being the object of worship; he worships alongside his people, leading their response. The Singing Savior demonstrates the appropriateness of musical and technical worship through multiple mediums including singing, prayer, and technology. In this way, since Christ is neither ashamed nor fearful, neither then should his brethren-children be.[4] Believers are at a place of honor because of the humanity of Christ, and thus technical artists craft praise of the Father through presenting Christ. Subsequently, Hebrews 2:12–13 can then be read to suggest that technology is more than just a tool; technical artists are themselves communication devices who connect God and his people as physical mediators of worship performed alongside the exalted Christ in their midst to bring the brethren-children together in praise of the Father. Technical artists take on the priestly role of pointing the people to their eternal High Priest in Christ, from whom reconciliation is found.

When technical artists use their art to produce a physical manifestation of psalms, hymns, and spiritual songs they become the dual vehicle for both the word of Christ dwelling with the congregation as well as the Spirit of God indwelling the people of the church. Janzen outlines the holistic unfolding of various artistic styles in practice:

> The hymn will satisfy our hunger for truth and depth of understanding and can help to express our reverential awe for God; the psalm will speak to our need for encounter and experience; and the spiritual song will stimulate the visionary in us and allow us to express personal feelings to God as our friend and companion.[5]

In this way, through their craft, technical artists minister to the people of God—elevating their spiritual maturity—because the process of performing teaches and admonishes by way of sound, light, and visuals. The imperative in Colossians is understood as present and continuing:

> Because participants are part of an ongoing creative process, either through improvisation or as a part of the community responding to the old as it is being created a new, they are actively engaged in the shaping of what may become a part of the communities' canon of

---

[4] There appears to be a two-fold insecurity held by church technical artists: (1) a sense of personal perfectionism causes a fear of failing to create a perfect worship service; and (2) many churches demand one hundred percent accuracy in moderating the weekly worship services—sometimes without the proper recourses to accomplish the job—which creates tension between the technical staff and church leadership.

[5] Janzen, *Psalms, Hymns, and Spiritual Songs*, 1435–38.

tradition.[6]

What technical artists do in practices defines the churches' liturgical and theological future exactly because it influences how the congregation both accepts the truths of the faith as well as their response to it. While worship production is multi-faceted, it has one directive: the gospel of Christ. Technical artists utilize the emotional outpouring of the heart found within the Psalter and cultural history, the musicality and artistry of Scripture found within hymnody, and the instincts for leading the atmospheric conditions through Spirit-led song. While psalms, hymns, and spiritual songs have a ministerial performance value within the sound, lighting, and video control, it is scripturally focused beliefs which should direct the worship. "Spiritual songs" suggest that the Holy Spirit can move the artist in a way that is not simply scriptural but also Christ-like, imprinting the message of who Christ is on the corporate body. Colossians 3:16 demonstrates that when a congregation gathers, it is not an individual experience of worship but an encouragement of one another corporately through the arts. Both the horizontal and vertical must be in mind when the church gathers. Worship music and production value are thus legitimate means when as a medium they point beyond themselves to focusing on the congregation for the purpose of exhorting and encouraging. There is therefore also an evangelical aspect involved: the final product also serves to educate and comfort the unchurched to become part of the church body. In this way, excellent biblically focused productions can help others endure trials by leading their worship in a way that allows them to experience the Spirit at work. Just as language about the power of the gospel is weaved throughout the letter to the Colossians through various literary styles, technically performed psalms, hymns, and spiritual songs provide a bridge between wrong living and a proper doctrinal understanding. Thus, technical artistry is not about the singing; it is about becoming stronger Christians through the worship experience. Technical artists ought to then be as equally aimed in Christ-like living as in the perfection of their craft in order to ensure the result is able to teach and admonish proper doctrine. They themselves serve a pastoral role in administering the Word of God. In this way, it would serve technical artists well to continually earnestly question whether each fader move, lighting cue, or camera shot is accenting the gospel message or not; is the focus pointing to the truth of Christ or simply show?

Today, those who work as tech leaders in all aspects of creative worship, such as audio mixing, advancing lyrical slides, shooting video, and controlling lighting, serve with Jesus as his mediators to the church, leading them in praising God. Fallen expressions become perfect acts of worship because Christ assists in the process as Perfecter of the faith. Thus, both Jesus and church technical artists serve a bi-directional role in mediating worship of the

---

[6] Hearon, "Music as a Medium," 186.

Father. Church technical artists specifically fulfill their role as they harness the power of their equipment, connecting stage to congregation to God in a way that brings them all into the presence of the Father through Christ. Just as Christ is the mediator between God and humanity, the church's technical artists are mediators between the stage and congregation because all created worship produced by the worship band and pulpit passes through the hands of the technical artists shaping the service. They translate how the action on stage will be received by the congregation. The sounds, lights, visuals, ambiance, and atmosphere resemble the role Christ plays in the text; they proclaim God to the people, creating a sanctuary for praise among the congregation back to the Father. All things are brought into unison: the audio, lyrics, lighting, and atmosphere are all to align through the actions of the technical artist in order to generate a distraction-free focus on the biblical truths presented. Worship moderated via sound and lights is also worship mediated by the living and exalted Christ.

Even though it should be apparent, note that it is not being stated that technical artists become "Christ" in any sense or that Jesus is the performer of the action. Rather, church technical artists serve in a similar role as that played by Christ in his leading of the worship between man and God. Produced music is "music in action."[7] It is doing something. Just as Christ is actively praising God and leading his people in praise, so too are the mediators within the church, the technical artists. Christ moved from atoning priest to interceding priest, leaving behind the earthly so that believers can connect to the Father through Christ as Mediator. Church artists moderate the way the word and worship are sent and received. They place the word into receptive action. In this way, artists' use of technology helps produce the convergence of heaven and earth through corporate worship.

As an example, Reggie Kidd notes that when a barbershop quartet sings in unity a "fifth voice" becomes present.[8] It is a voice that encompasses each individual but is greater still than the sum of them all. When singing is performed in unison with Christ, the fifth voice who is Christ is manifested within the midst of the congregation.[9] When technical artists create the musical worship mix, Jesus's voice is unified with the church's. Worship becomes Christ's encouragement and is heard as perfect by the Father. This is further supported by the book of Hebrews, in which the author calls Christ, the Worship "Liturgist" (*leitourgos*) in 8:2. What God desires from worship, he provides through Christ. The effect is that corporate worship practice

---

[7] Begbie, *Resounding Truth*, 60.

[8] Kidd, *With One Voice*, 12–13.

[9] Christians often cite Matt 18:20 as a proof-text of Christ's presence when two or more believers gather in his name. This is not a direct cross-reference; however, it does demonstrate that Christ's presence joining among his people is a common belief.

connects worshipers with the worshiped.[10] Favor and acceptance before God are not through perfect worship but through perfected worship because of a unified response that is mediated to the Father through Christ's saving work. With certain aspects of a church service being horizontal in that they serve the people, worship of God himself is vertical; acts of worshiping bi-directionally connect God to man through Christ the Mediator, and the people to one another through Christ the Brethren-Son. Christ's dual identity becomes the model for the church technical artists' purpose in performing their craft. Thus, the message of Christ should be at the center of all church tech activities. The production itself is secondary; it is the medium for the biblical message being delivered. What technical artists produce, they minister.

THE CREATION OF SACRED SPACE: "TENT" BUILDING

One of the more recognizable roles modern church technical artists fill is that they aid in creating and defining the sacred space from which the congregation worships. Today, communication is faster and more effective than ever before, allowing instant access to the world and one another at the touch of a button. Technology brings people together, building *koinonia*. As such, worship can be judged by how the church community is brought together in unifying its technological voice. The engagement—or lack thereof—creates its identity. In the United States, it is common for churches to meet in business parks, high schools, performance theaters, and other secular locations which are transformed into the worship space for the weekend service. Technical artists hold a central role in converting the place into the usable sacred space. Eugene Peterson sums up a ministry meeting with an architect for the construction of his new church by stating, "there was more to church than a building. We needed a building. But we were not about to be reduced to a building"[11] Peterson noticed that the designer was so focused on what the aesthetics and functions would be that he failed to ask about the identity of the church itself. In this same way, the Bezalel passage is more than a construction narrative. The tabernacle is a "tent of meeting" (*'ohel mo'ed*). It is the physical location where the Israelites can commune with God in their midst. Hebrews refers to Christ as the "true tent" (8:2), which complements the OT citations from David and Isaiah in 2:12–13—placed by the author as Jesus's words—noting that it is through trusting in the Lord alone where refuge is found. In Colossians, Paul's concern is creating a sacred space through

---

[10] Though not expressed in this same way, the Second Vatican Council likewise declared that Christ's connection brings the heavenly to earth, and the songs that are sung are those which are heavenly, yet connected through the earthly church, which in turn is performing Christ's work in the world.

[11] Peterson, "The Pastor," 90.

holistic multi-medium worship practices that build up the corporate body. Thus, the performance of psalms, hymns, and spiritual songs creates the social and religious space for the modern church and one method for how believers can connect to and communicate with Jesus today: through worship.

For the Israelites, aesthetics played a significant role in skilled work when produced for the public sphere because it served as an outward demonstration of their faith in God alongside their acknowledgment of the centrality of God. The labor for the tabernacle is inherently a social practice. The materials are offered up voluntarily by the people, and the work is executed voluntarily by those whose hearts are stirred by God. It is communal in nature: by and for the people of God. This sits in sharp contrast to the forced labor that opens the book of Exodus. The tabernacle text, therefore, results in a corporate understanding of the worship of God. Just as God could have built the tabernacle himself—like inscribing the stone tablets—he chose instead to use worker-artisans, personally called by name, so that the artists could take ownership of their creation, honoring the final product as the sacred dwelling space of God that they took part in producing. In this way, Exodus is a metaphor for constructing the church through skills gifted from God and harnessed through the Spirit, in order to build the sacred space for the worship of God. Bezalel and Oholiab, who are leaders and members of the congregation, are moved by God for the benefit of the people.

The book of Hebrews presents Christ as the Singing Savior; what he sings is not a call to forge a new physical building for him to dwell in but to turn those who believe in him into living buildings. While the church is God's people, the true tent is not a place but a person. Christ is the Architect of the house of praise, believers are his house, and Jesus confidently boasts of them. Auditory and visual worship done through Christ brings the people of God together in one accord because the horizontal expression is experienced through the vertical Liturgist. Endurance and dependence on God instruct earthly production techniques, mediated between stage and congregation.

The purpose of the early church's performance of psalms, hymns, and spiritual songs is to express the church's corporate faith, which in turn forms the foundation for liturgical practices. Worship promotes growth in the knowledge of who Christ is and what he has done for his people in redemption. It is both an emotional and intellectual endeavor. It is "an ecclesio-centric aesthetic."[12] Holistic worship practices create a unified body of all ages, sexes, and societal statuses that instruct the church to put-on qualities of Christ and put-off former habits. In this way, "tent building" is viewed as successful when the church unifies through the technical arts.[13]

---

[12] Harris, *Curious Works of Bezalel*, 14.

[13] Even though unrelated, it is noteworthy that by trade Paul was a tent builder. He himself moves from builder of dwelling places to the physical planting of churches to the building of God's people.

Even though it is common to compare the tabernacle to the church, where the tabernacle is simply a tent-building, the NT church is neither a tent nor a building, but a people. Thus, technical artists today perform their tasks through God, for the benefit of their congregation, in order to connect the people to God and one another. For example, popular worship leader Kari Jobe noted an experience where the technical artists created an immersive experience called "environmental projection" where various names of God flooded the walls, scrolling from floor-to-ceiling, left-to-right as the bridge to "How Great Is Our God" began and the lyrics "Name above all names" were sung. Jobe explains:

> You should have seen everyone gasp for air. . . . It was one of the coolest worship moments I remember ever having. It took all focus off the music, off even just what we were trying to do to connect with people, and it just was all about God."[14]

Brian Certain, creative director at FUMC Mansfield, explains the reaction from his audience, "[They] say for the very first time, they 'get' technology in worship. It is about taking the technology that God has gifted someone to create and taking that into a very reverent and traditional way."[15] In that technologically articulated corporate experience, "when words and music are combined as song, the result is a distinctive communicative medium that is neither wholly words nor wholly music,"[16] nor wholly technological. Sacred space in the form of art becomes a tangible, experiential vehicle for God's revelation of himself to the believer.

BEAUTY: AESTHETICALLY PLEASING TECHNICAL ARTISTRY FOSTERS CULTURAL HOLINESS

Artistry is customarily defined and judged by its aesthetic qualities. Technical artistry is no different: sounds, visuals, and environmental experience form the congregational opinions of the worship service's beauty. "All too often Christians settle for something that is functional but not beautiful."[17] High art is something special and to be desired, not despised. God bestows the abilities on a few, and even though it is possible for it to be abused, when performed in accordance with the Spirit, it brings honor and glory to God. Art is within God's will. Technical art, the text suggests, ought to elevate sacred space for the advancement of God's kingdom, reminding the people of the magnificence of God, and offers a glimpse into the heavenly realm.

---

[14] Cameron Ware, "Environmental Projection," *VisualWorshiper.com*, 2012. htttp://www.visualworshiper.com.

[15] Ibid.

[16] Hearon, "Music as a Medium," 181.

[17] Ryken, *Exodus*, 946.

For technical artists, it is not about the gear, but the method. Throughout the three passages, beauty and excellence are continually weaved together alongside the materials as complementary and necessary elements to produced and performed worship. In the tabernacle construction account, God commands that his dwelling be built from the best materials, including gold, silver, bronze, fine linens, and precious gems. The narrative presents artistic and creative abilities as a gift of grace from God crafted together with wisdom, intellect, and learned skills. Hebrews declares that Christ is present in the worship experience. Nicolas of Cusa (1401–1464 CE) went so far as to suggest that Jesus is the "Art" of God, present in our worship. Even though this description could be viewed as problematic if it were to speak of Christ as being "created," the suggestion here is that it demonstrates God using a "person" for his purposes, which is in God's nature to create what is perfect and true for humanity to mirror. Colossians presents three distinct mediums for worship, all of which carry their own characteristics of beauty. The psalms are OT or historical songs ingrained with musical, liturgical, and literary performance directives; hymns are thankful praises to God and his characteristics; spiritual songs are outbursts of the Spirit that demonstrate the heart-transformed position of the worshiper. All three possess a sense of being naturally beautiful because they are artistic interpretations of the church's beliefs about Christ, God's "Art." Today, people have a hard time thinking of God as beauty because they view beauty in solely aesthetic terms, like color, shape, and sound. Yet, biblical beauty is holistic, incorporating the good of the creation and the work of God in salvation, including making all things holy. God finds beautiful that which he makes holy.

Leading up to the tabernacle construction narrative, the author cites the high priest's garments as embroidered with "dedicated to the LORD" (Exod 28:36). Likewise, the purpose is noted as "for glory and for beauty" (Exod 28:39).[18] In this way, the text suggests that created works used for worshiping God are not only to be beautiful but also must bring glory to him: God is to be glorified in the creation. "God, the designer and maker of the universe, clearly places great value on details of design, construction, and artifice."[19] This gives a clue as to the purpose of the tabernacle artistry: its beauty points to its holiness. Its beauty is visible through forming only the finest materials because the materials serve as metaphors for the "quality" of God: what is valuable must also be valued and possess greater holiness. In this way, Bezalel and Oholiab are akin to Michelangelo and da Vinci, who worked in various mediums, able to shape the finest forms out of the finest materials. In a society such as developmental Israel, as opposed to neighboring Egypt or Mesopotamia, art had to set a practical purpose. Art was never created simply for

---

[18] The word for beauty, *tip 'äret*, is translated here as beauty or ornament but is also translated elsewhere as splendor, glory, or radiance. In the LXX it is the word *timén*, which has the sense of honor and value.
[19] Veith, *State of the Arts*, 106; Veith, *Gift of Art*, 19.

art's sake. In fact, the tabernacle includes great works of art that would only be seen once a year and by only one person, like the ark within the holy of holies.[20] In the tabernacle there is no distinction between beautiful and useful; every object contains both properties. And therefore, what is useful for the practice of worshiping God ought to also be made beautiful. In this way, the imagery Bezalel presents implies that a sound biblical truth can be wrapped in the beauty of gold, silver, and acacia wood, as well as camera angles, aesthetically pleasing lighting, and feedback-free audio.

Hebrews 2:12–13 offers a bi-directional Christology of creative worship wherein Christ is present with creatives in their productive acts. There is no need to invite Christ into worship because he never left. He is present in and through the sound booth and lighting console. In an artistic sense, a high Christology demands that only well-trained creative capacities be used in praising the Creator, which involves personal attitudes, acts of praise, and corporate adoration in response to God and his saving suffering. Produced worship is not excellent because of anything the individual has done but because of the high priestly, apostolic work of Christ. Scripturally rich and biblically founded acts of creativity that mirror the identity of Christ are to be expected. Effective worship mirrors what is witnessed in Christ: the singing, praise, and proclamation of worship to God. Christ is active because it is his ministry; his children are his ministry, and they have a role to play. Praising his name requires presenting worship that is of him and to him, beautifully and pleasingly expressing the qualities of who he is and what he has done for the worshiper. Even art embracing difficult themes, like the Passion, can be presented beautifully when the technological message highlights the work and purpose of Christ as the key participant for his brethren-children.

Even though John Calvin proclaimed that Jesus is the Chief Composer of our hymns, the Reformed tradition often usurps the importance of worship for solely Word-centered preaching. Often in contemporary practice, hymns are considered important sources of expressions about God, while the act of worship is often positioned as secondary to preaching. Colossians suggests that a multitude of worship is encouraged, accepted, and serves as an equal and valid method for teaching God's people. The text suggests that modern church production ought to have a rich variety of mediums, visuals, moods, and styles. Technical artists are to create an immersive experience: some joyful expressions, some praise, some thanksgiving, some of profound biblical doctrine, some dark, some bright, some loud, some soft, some acoustic, some colorful, and some plain: it does not have to be the same thing over and

---

[20] This theology contends that this common thought is a misnomer. The "people" would also see the ark as well as the inside of the holy of holies on a regular basis during the construction and deconstruction as the Israelites moved throughout the desert. A better understanding would be that the high priest is the only one who would see or use these in religious practice.

over.[21] Beauty is created by expressing the realities of the faith and of Christ. In a practical sense, church technical artists have the freedom to step out of their comfort zone of routine to produce a balanced approach of a blending traditional with modern, being directed through wise, trained knowledge of contemporary cultural practices. Technical artists are able to re-write the methods for engaging with God by mirroring the same methods the congregation interacts with in their secular lives. For example, concert style lighting can connect the congregation to God when used to create moods complimentary to the vocals and projected lyrics. In this way, God is presented as beauty, usurping the medium for the message. As David Pao suggests: "more theology is engrained into our hearts through singing than through the printed page or even through preaching."[22] The more technical artists effectively embody psalms, hymns, and spiritual songs through technological practice, the more the body of Christ grows in their knowledge of Christ.

In a practical sense, technical artists have a direct impact on mood, engagement, cultural understanding, and holiness of the worshiping congregation through the beauty they create—both positively and negatively. Jim Kumorek and Duke DeJong offer two tangible examples of this. First, Kumorek's case study on church lighting finds:

> Whether you're in an old cathedral, a traditional sanctuary, or a modern auditorium, how you light your church service impacts the mood and how well people will comprehend and retain what takes place in your service. . . . Lighting serves several functions. First . . . is to simply make the platform and the people on it visible. . . . When lighting is too dim . . . this cuts down the amount of brain-power that they will put towards paying attention to the actual service, as they are spending time locating those who are speaking. It also is very tiring, and will cause people to start tuning out or disconnecting. . . . Conversely, lighting that is too bright can be painful to look at, and also causes people to disconnect or stop watching. Another purpose of lighting is to support and enhance the mood of what's taking place in the room. . . . Colors impact one's mood, and while your goal isn't to be manipulative in a negative way, colors can help create an environment where the natural emotions that worship can bring out are supported and reinforced. Brighter colors such as reds and yellows help reflect energy and excitement; colors such as deep blues and dark greens reinforce introspection and peacefulness.[23]

---

[21] This brings up and interesting debate regarding structured liturgical practices. To what extent is variety encouraged and demanded? Is it solely in style? Does it extend to the order of service? There is a much scholarship in biblical worship circles in regards to proper service structure and maintaining tradition. The text appears to suggest that service order and style is less important to the overall worship service as creating a holistic experience that is both aesthetically pleasing and Scripturally rich.

[22] Pao, *Colossians and Philemon*, 248.

[23] Jim Kumorek, "Case Study: The Pastor's Guide to Lighting," *WorshipFacilities.com*, June 15, 2016, https://www.worshipfacilities.com/gear/case-study-pastors-guide-lighting.

Next, industry consultant and former technical director Duke DeJong writes about the use of EQs in audio mixing:

> Just as each person's voice has a unique makeup and signature, every instrument or vocals has a makeup of frequencies that is unique to it. . . . Our most critical source in the church, the human voice also has clear-cut frequency ranges, regardless of whether your voices are singing or speaking. . . . The most important frequency range in the voice in my opinion, and the one I see most commonly mis-adjusted, is the intelligibility range in the high-mids (2 kHz to 4 kHz). . . . When listening to vocals that are "honky" or "tinny," I often see sound guys reach for the high-mids and adjust those down to try and improve the sound. . . . [Doing this,] we're actually attacking the intelligibility when lowering the high-mids, and missing the "honky/tinny" sound that's in the 400–2 kHz range. It seems like such a small miss on paper, but this mistake will often cost the vocals their clarity in the mix. . . . Our vocals and instruments are living, breathing, unique things and they all have their own flavor.[24]

Thus, lighting and EQs are able to cut and add, controlling the clarity and effectiveness of the overall beauty of the mix because vocals, instrumentation, lighting, and environment are not static but organic and dynamic. The words alone in preaching and song are not enough. Through mastering their visual and auditory tools, technical artists can make generic mediums and make them holy through manipulating the perceived level of beauty. How they do it generates a tangible effect on the congregation's understanding of the theology by impacting its intelligibility. When done poorly, the product risks becoming a distraction, causing the worshiper to tune out the theology presented in order to focus on the distraction. Thus, the beauty and aesthetic are directly tied to the congregation's ability to become holy. In this way, what is holy is also presented as wholly beautiful.

INTENTIONAL WORKMANSHIP:
PLANNED AND PURPOSEFUL CREATIONS

The process of creating music influences the person. The amount is debatable, but recent research suggests that a psychological and physiological change does indeed occur. Art possesses a force within itself. Believers hold certain truths about God, and when those truths are presented through technological illustrations, worshipers are able to express their feelings about those truths better. Artisans add an emotional layer, which allows the state of

---

[24] Duke DeJong, "Church Sound: Mixing Like a Pro, Part 4—Making EQ Work for You," *ProSoundWeb.com*, December 17, 2013, prosoundweb.com/channels/church/church_sound_mixing_like_a_pro_part_4making_eq_work_for_you/.

the heart to be presented. In this way, musical composition, production value, and environmental presentation can influence both the individual's and corporate church's emotions and psyche. Thus, technical artists define the church culture. Their worldview, emotions, visions, and values about the world around them become the expression of the congregation for which their art is created. "What is happening in the arts today is prophetic of what will happen in our culture tomorrow."[25] The way church techs use technology can either shape or will be shaped by society, and the techniques they use will have a tangible effect on the way the congregation experiences the worship service, and therefore God. Exodus, Hebrews, and Colossians together can be read to support the idea that because technical artists can influence the beliefs of the congregation, how they perform their craft is to be intentional and purposeful. In Exodus, God commands that craftsmanship, skill, intelligence, and wisdom are all necessary elements for creating the people's place of worship. They are to master their craft in a way that is vocational and not haphazard or amateur. In Hebrews 2:12–13 the author presents the congregation joining together with Christ in worship through the revelation of God to the people and the people's response to God. It is a decisive recognition of bi-directional worship among the people and toward God. What the people sing in unison is how they present themselves to the Father. For the Colossian people, artistry is a multi-medium worship experience of traditional psalms and cultural history, hymns of praise to Jesus, and spontaneous spiritual songs sung out to God. The corporate act of worship edifies one another. Worship is not simply musical but theological: substance trumps style. When Christians teach and admonish one another through worship, life change happens because worship promotes the indwelling of the word of God among the people.

Exodus's listing of the five necessary non-spiritual qualities—skill, intelligence, knowledge, craftsmanship, and teaching ability—can be read to suggest that artistry, as presented in the Bible differs greatly from how artists are commonly thought of in modern culture today, where they are often considered to possess a humanistic talent of rare genius. As a gift of grace from God, all artistry for the worship of God is to be intentional, purposeful, and planned. It starts from contextualized blueprints and is skillfully molded into a work of intricate art.

> Just as prevails today in God's work, some skills are the result of a lifetime of study, training, and experience, some are the result of special divine intervention and guidance, and some are learned from others in the process of carrying out a given assignment for the Lord.[26]

For the technical artist, it is a combination of them all. In modern society,

---

[25] Ryken, *Exodus*, 946.
[26] Stuart, *Exodus*, 759.

objects of aesthetic senses and artistic creations that take hard work are often overlooked. For example, the architect who has the ideas is viewed greater than the skilled laborer who can turn the plans into reality. Likewise, the church technical artist who advances the lyrics slides is seemingly viewed as less than the vocalists on stage singing the words. The lyrics advancer, however, plays a special roll in feeding the gospel truth to the congregation with the power to impact the delivery for better or worse. Thus, the call to create is not solely for the sake of making something that looks nice but to bring honor, glory, and majesty to the recipient, which is God and his people.

Theologically, the Decalogue is not against creating art; it is only against art created and worshiped as God himself. Historically, faulty exegesis of Exodus has led the evangelical church to abandon the arts vocationally. Yet, God is not against beautiful creations, only objects that are worshiped in his place like the golden calf. There were distinct times in the history of church art that it was to be destroyed, like during the era of the Iconoclasts. However, they were not always opposed to the use of art, just the abuse of it, as witnessed through commissioning great works of art during the same time period like that of the Sistine Chapel. Thus, humans are not just made for work (Gen 2:15), but artistic work. In fact, Exodus 35:30–36:1 not only shows that humans were made for artistic work, but work of high quality and aesthetics. This work is a purposeful and holistic encompassing of the five gifted qualities aligned with the Spirit of God.

Creative ability is affirmed and fulfilled through Christ as a form of worship. According to Hebrews 2:12–13, arguments about style are problematic because Christ accepts all his brethren-children's actions performed to and through him. The substance of worship supersedes the style of worship because technical artists who perform their craft according to the identity of Christ have their productions perfected by Christ, whether acoustic, rock and roll, hip-hop, candlelit, concert lighting, written word, or video. Worship has an active dynamic at work; it is not that worship is performed, but that it is perfected in connecting the people to God and God to the people. A church technical artist's creative worship is not accepted because of his or her work, but Jesus's. The artistic medium becomes the message of Christ. As technical artists endure to skillfully perfect their abilities, performing their craft in praise of God's name, the congregation is unified in Christ.

Colossians 3:16 falls within the ethical conduct section of the epistle—often referred to as the "put-off put-on" passage. It focusses on what to do. The technical performance of psalms, hymns, and spiritual songs cannot be haphazard but rather must be presented clean and straightforward so that there is no ambiguity. The affect is instructing the congregation of the tenets of the faith and person of Christ. It must be intentional and complementary to the lyrics and style of worship. Unintelligible worship would thus have no purpose outside of oneself.[27] The technical arts would then be complementary to the worship experience and not the attraction itself. As a practical example, technical arts director Jose David Irizarry writes regarding the layering of the band and vocal mix:

> Vocals are the crown jewel of the song. You can only do so much by riding the fader. For excellence and professionalism, your mix requires more preparation, knowledge and skill. . . . In general, bass and drums are the cornerstone of a musical theme in a band. Then guitars, keyboards and other instruments complement the harmonic setting of the musical arrangement. Finally, on top of all this, vocals. . . . Pay special attention to the lead vocal mic. The lead vocal or worship leader has to be clearly distinguished without overpowering the others. The lyrics, as well as any spoken words during the performance, need to be heard clearly by the audience. The BGV and/or choir level must fit in the whole mix.[28]

Thus, for worship to be useful for teaching and admonishing, the production experience must be understandable; it must be "fitting." How the elements are fashioned together matters. Layering in audio is not simply for aesthetic effect but gospel intelligibility. Technical creations cannot be frivolous; they must exhibit truths about who God is. They must be uplifting and adapted to instruct the whole corporate body.

Because the creation of art is itself a biblically sound vocation, being intentional in workmanship implies that those involved in the production are not amateur in ability, but professional.[29] The modern term "vocation" can be equated with the biblical word used for "calling." While a career is something chosen and compensated for, a calling is responding to a personal selection by God for a particular role. Thus, the view presented here is that church technical arts can be regarded as a Christ-sanctioned vocation because artistic knowledge is a learned and performed skilled aligned with one's personal calling.

The elements of the craft, like light, sound, video, and environmental control generate the mood with which the congregation will engage. That mood defines the way the congregant-customer views the object of the craft, which in the case of church technical arts is God. When the lighting complements the stage activity, greater audience engagement happens; the result is greater

---

[27] In 1 Cor 14, Paul unpacks the necessity for intelligibility in worship. He writes that those who speak in tongues are not speaking to man, but to God (v.3). When the word is preached, it must be easily understood by the audience (v.9), so that the end result is a building up of one another (v.12).

[28] Jose David Irizarry, "Vocal Mixing for Live Events: Proper Approaches to Getting It Right at the Source," *ProSoundWeb.com*, September 6, 2013, https://www.prosoundweb.com/channels/church/church_sound_the_main_aspects_of_vocal_mixing/.

[29] Volunteers in the church are important, and it is how most churches run their tech ministries, but this is different than whether or not the church tech is compensated for the work performed. Being professional is a standard of quality delivered, not the employment status of the technician. It denotes a continual development of one's craft due to a particular calling to serve.

dwelling of the message of God. In the same way, the audio mix is able to elevate, highlight, and hide specific frequencies and sounds, just as the preached message can highlight specific beliefs of the religion. For example, the use of vocal layering in the audio mix can fortify the delivery of those beliefs. Placing the vocals (i.e., the presentation of gospel) "on top of" the production mix demands that what is being highlighted be theologically and doctrinally accurate and that the mix be clear and dominant. The skills to effectively produce biblically informed worship are the result of continually perfecting the craft and increasing in the knowledge of Christ.

POSSIBLE OBJECTIONS
TO THIS CONSTRUCTIVE THEOLOGY

As the first to explore the theological relationship between the technical artists and a biblically informed praxis, it is not feasible to cover all possible scriptural references pertaining to it, arguments for or against it, nor answer all objections. Certainly, some findings could be challenged.

This book goes against the modern idea that artistry is merely the combination of skill and inspiration. In 2003, famed secular scholar, teacher, and artist Eric Booth sought to form an inclusive definition of a "teaching artist"—a term that would firmly fit the role of the technical artist as described. He suggests that a teaching artist must: (1) consider the audience; (2) model through embodying both the teaching and artistic process, incorporating various pedagogical techniques; (3) be a professional in the craft by focusing on the process in addition to the results; and (4) be aware of his or her dual role as both creator/teacher and participant.[30] According to these findings, the church technical artist would completely fulfill his or her role simply by possessing the skills, wisdom, and ability to embody the art as it is passed along to the next generation. Even though non-believers could create satisfactory productions in the context of a worship service, this theology of church technical arts argues that only an artist who is a believer can present the production in a way that also speaks to the spiritual needs and beliefs of the congregation. Booth adds that "teaching artists connect their art form to other important areas of life, . . . [and] relevant aspects of people's lives."[31] Though Booth downplays it, this is indeed a necessary addition. Technically based worship would demand that the artisan is emotionally and spiritually invested in the spiritual development of the recipient. The bodily effect of being filled with the Spirit allows them, in worship, to offer themselves authentically, creating worship that does not need to be perfect but rather simply in

---

[30] Booth, "Seeking Definition," 7–10.
[31] Ibid., 7.

accordance with the person and work of Christ. In this way, it can be suggested that a church technical artist must also possess an indwelling of the Holy Spirit in order to holistically fulfill the calling as a "teaching artist." Where excellence is an expected standard, so too would be "being a Christian." In practice, when combined with environmental additions of sight and sound, the experience presented to the congregation can mirror the presence of the Spirit among the people. Thus, this theology argues against certain practices like hiring non-Christians as musicians and technical engineers. Even though these artists could create pleasing sounds, artistry in the church requires that the performers and producers are to be believing worshipers themselves. While many could argue that allowing non-Christians into the tech booth serves an evangelical purpose—creating opportunities for the Spirit to work within them, eventually leading them to salvation—and though the practice is common and may hold merit, it is not present in the texts as explored here.

It could be possible to reject these findings because the NT church worships "in spirit and truth" (John 4:24)[32] now that the glory of the Lord has left the physical temple (Ezek 10:18). Indeed, commanded, structured worship practices are no longer required elements for NT Christians. Even Richard Viladesau writes, "Christian worship should aspire to the philosophical idea of 'spiritual sacrifice,'"[33] and Eugene Veith suggests that modern readers should not to make too much of the tabernacle because it is no longer needed now that believers worship in spirit and truth.[34] If they are correct that we should not make more out of it than a "philosophical" understanding, then the conclusions drawn here defining a technical artist fall short. Indeed, the Spirit is an important aspect of modern Christian worship because at the core is a working of the Spirit within the artisan. While the tabernacle is not the modern church—since the church is its people and not a building—the tabernacle narrative can be used as a model for modern practice specifically because of the Spirit at work. If Christians do worship "in 'Spirit' and truth," then biblical texts that present the action of the Spirit ought to be given special attention. The findings here show the Spirit either generates the tangible actions or perfects the qualities the artist is to possess, and all are done to and through Christ-revealed. The technical artist is not merely a passing medium through which worship is performed. God's glory leaving the temple allows for the worship of God not in any one location where God commands but any place his people gather. Therefore, the passages suggest there is a particular duty for technical artists as mediators of the faith to produce worship

---

[32] It is important to recognize that various translations differ on the reference to these words. Only a few (CSB and NIV) capitalize "Spirit" making it reference the Holy Spirit. None capitalize "truth," suggesting it is not a reference to God as Truth, but rather true statements about the faith.

[33] Viladesau, *Theology and the Arts*, 16.

[34] Veith, *State of the Arts*, 113.

due to and through the indwelling of the Spirit among the people of faith within their common place of assembly.

Next, it could be argued that this book is artificially manufactured to include findings in the affirmative. As a practitioner in the technical arts industry, I could be viewed as biased toward the conclusions presented. To counter this concern, "warning" arguments are presented to the proposed chapter findings. In light of the golden calf narrative, chapter 3 suggests that the act of producing worship is subject to idolizing the creation process and product, generating a roadblock to the congregation's worship of God. Chapter 4 contends the conclusion that technical artists serve as mediators between God and congregation could be flawed because—as the author of Hebrews maintains—all human mediation is insufficient as viewed by Aaron and the high priest's inability to wholly atone for sin; the only qualified mediator is Christ himself. Lastly, chapter 5 suggests that similar to the concerns of the contemporary "worship war" battles, the Colossians text warns against the heresy of embracing popular culture which can lead to worship being entertainment rather than teaching and admonishing the congregation. It is important to note that the research question explored in this book answers how a biblically informed theological understanding might inform praxis; the intent was not to establish a special status for technical artists. For, indeed, the Bible does not specifically speak to technology, the technical arts, or church technical artists. When performed accurately, a proper biblical exegesis results in neither positive or negative findings, but rather offers an examination of the biblical text understandable for the present audience. Examples of the technical arts in practice are offered not as justification but tangible examples of the principles presented; likewise for the examples offered within the warning sections. Yet, the warnings sections demonstrate that the application of the technical arts to praxis can also make a negative impact on the worship experience. Nevertheless, the all three passages present inherent warnings for practice that could be viewed as challenges to the application in practice, areas that contemporary technical artists must be aware of when in the midst of their craft.

Last, the selection of the three passages could be viewed as too selective or incomplete, creating a portrait of the modern church technical artist that is either too narrow or not fully informed. Yet, the selection was not haphazard as noted in chapter 1; each passage individually had some amount of research within the context of worship or creativity, even though not wholly explored in that light as a scholarly consensus. This study specifically chose not to look for themes around technical artistry in standard artistic passages like the Psalter. For this reason, no cross-references to other verses to "fill in the gaps" were presented; but rather, biblical passages are cited if they are dependent parallels. This theology allows the scholars and texts speak for themselves, regardless of the outcome. In fact, it is quite astonishing how well the findings inform and support one another. This reveals an important point that technical artistry is a complete embodiment of the person: artist, technician, theologian, Christian, worshiper.

CONCLUSIONS: PLANTING THE FLAG

The view presented here places the technical arts into the theological school of ecclesiology. The technical arts are commonly aligned with "worship" or "musical aesthetics." However, this book suggests the technical arts are more than merely the creation of visual and auditory pleasing aesthetics but are ecclesiological in a complete sense: what technical artists do and how they do it influences the church body. Church techs engage the worship service holistically. Technical artists are "members of the band" while also being essential in the presentation of the preached word; their job does not end when the musicians exit the stage. They create a sacred space from the moment the congregation enters the worship center until they exit. How they perform their craft has the ability to impact the spiritual development of the congregation beyond the momentary worship experience.[35] They have the ability to emphasize the theological points buried within song lyrics and sermon points through audio dynamics and visual light and shade.

This theology offers the view that technical artists are themselves mediators of church worship. Where the pastors and worship band are the presenters of the preached and musical word, church technical artists translate that information to the congregation. How church techs perform their craft can shape the congregation's beliefs about the religion and provoke both an emotional and physiological effect, which ultimately impacts how participants engage with God, one another, and the church service itself. Where Christ serves as the mediator between God and humanity, technical artists mimic Christ's role as mediators of the electronic signal from the source (pastor's pulpit microphone, worship leader's microphone, guitar amp, piano, etc.) to the tech booth where technical artists manipulate the signal, interpreting it for the appropriate output to the loudspeakers and projection screens. What the congregation receives is not the initial source, but the source mediated through the technical artist's interpretation for best results. In this way, church technical artists are architects of how the church relates to its sacred space through influencing environmental stimuli.

---

[35] Technical artists can continue to instruct the congregation throughout the following week by creating artistic mediums that implant in the hearts and minds of the worshiper. One example is producing emotionally charged and theologically rich sermon bumper videos that psychologically and spiritually shape the congregation both during the service and after through being shared throughout social media platforms.

This book aligns the technical arts with the long-standing tradition that the musical arts are a worship-centered activity and that those who create technical productions are worshipers in their own right. The defining characteristic between a secular artist and a church technical artist is the indwelling of the Spirit. This constructive theology presents the view that possessing skill, ability, wisdom, and knowledge of craft is not sufficient; a church technical artist must also be filled with the Spirit of God. Spirit-led artistry means the technician's identity is as a worshiper of God. It is not confined to the momentary occurrence of producing the worship experience but is the defining factor in a life aimed at becoming Christ-like.

The view presented is an alternative reading of the phrase "psalms, hymns, and spiritual songs" found in Colossians 3:16 and Ephesians 5:19 in the context of Paul's exhortation to teach and admonish one another through the act of worship. Where it is common in most of the Pauline studies to assume a Jewish reading of the text, the Colossian context is clearly Greek. A traditional view of Pauline writings being foundationally Jewish would not fully apply in this case. Therefore, a Greek understanding ought to additionally be considered, which includes a Hellenistic interpretation of the terms individually. In Ephesians Paul builds upon the phrase, by adding a Jewish flavor. A contemporary reading, then, possesses both Greco-Roman and Jewish connotations, which offers a broader meaning to the terms.[36] In this way, a wide variety of technologically produced worship techniques, styles, and mediums are to be encouraged and pursued because all are able to speak to and lift up the corporate body.[37]

Technical artists are to be scholars of theology in their own right in order to be fully effective technicians for use in the church context. The ability to execute their craft extends beyond the capacity to complete their tasks; it requires an understanding of how their actions align with the biblical

---

[36] For example, psalms are not solely the OT Psalter but also include the meaning of "plucked instruments;" therefore, not only scriptural but musical methodologies are effective mediums to teach and admonish. Hymns are not solely praises to God but are also festive songs of high poetic and stylistic craftsmanship. Spiritual songs are not solely songs about God but are also artistic works inspired by God and performed due to the Spirit of God working in and through the artist.

[37] This study rights what I believe is a popular misconception in Pauline studies about the understanding of worship practice in Colossians and Ephesians. I contend that most theologians and scholars use a post-temple destruction understanding—more aligned with what would come about in the second and third centuries—rather than a pre-destruction reading. The epistles were both penned in the mid-to-late 60s, prior to temple destruction and, therefore, Paul and his audience could have only read the text according to religious and societal practice at that time. Even though the temple was destroyed shortly thereafter, the meaning when written and delivered would have been more holistic than expressed in modern commentaries that suggest post-temple destruction practices. Scholars commonly state that worship would have been solely lyrical and not musical. Yet, temple practices which would have practiced in local synagogues pre-70 CE were instrumental in nature, containing a broad range of musical and poetic styles, from lament to praise. Practices would include hymnal or simplistic vocal chanting as well as celebratory worship. In modern application, this view suggests that worship is to be a holistic practice from traditional to spontaneous, from text based to heart based, and from vocal to instrumental to technological.

narrative and their place in God's theodrama. Adjoining theology to practice is to allow the biblical text to inform praxis alongside mastering one's technical ability. The substance of the message trumps the style of the created performance. In fact, the Bible affirms church technical arts as a biblically sanctioned vocation. Those who work in the church context are responding to a personal call that is from God and a response of the Spirit within them. Their artistic calling is a Spirit-led vocation that requires a response of professionalism and high craftsmanship. Skill, knowledge, and wisdom are the result of lifelong dedication to one's craft. Being ecclesiological, the technical arts are a biblically sanctioned church department in which investment in human and capital resources would be encouraged. Likewise, to be ecclesiological, technical artists must be learned themselves in both theology and church doctrine in order to identify how their choices impact the church itself.

Modern church technical artists are to embody all six qualities associated with Bezalel and Oholiab—Spirit, skill, intelligence, knowledge, craftsmanship, and teaching ability—in order to create a sacred space acceptable to God. Even though great art can be created without possessing all six, a holistic approach to the creation of the church's sacred space requires all be balanced to be effective. Additionally, incorporating all three passages exegeted, this study argues that technical artistry is a comprehensive enterprise that encompasses ability, craftsmanship, worship, community, wisdom and instruction, heart, artistry, multi-medium mastery, teaching ability, edification, communication skills, and the Holy Spirit. While each quality can individually create pleasing art, the biblical narrative suggests that pure aesthetics in terms of church production is not its telos. Production-based technical artistry is not solely a work of the head, hand, or heart; it is an embodiment of them all working together in unison, serving as a communication device of revelation and response that connects God to his people, the people to God, and the people to one another.

THE NEXT STEPS

This constructive theology serves as the first theological doctoral-level study on the church technical arts, which until now was a void in biblical studies, which scholars like Frank Burch Brown referred to as the "blind spot" in the scholarly literature. This study opens the door for scholars to consider how the use of technical liturgy fits into worship and ecclesial studies. It specifically focuses on a biblically informed understanding of technical arts in praxis and connects traditional production arts to a twenty-first-century context. It establishes why, biblically, the technical arts impact the delivery of the message and influence the congregation: it does so in the same way that vocal music, instrumentation, paintings, sculptures, and stained glass were utilized historically. Examples of technical artists in practice spread throughout the book are offered here to demonstrate how biblical principles can shape practice. This book serves as a springboard for further exploration in the relationship between church technical arts and the scholarly disciplines of biblical studies, practical theology, and worship studies. Indeed, many avenues for future research exist.

One direction for future research is how this study is impacted by non-believers (or even those of other faiths). Surely, excellent technical works are possible for the creation of sacred space by those not of the Christian faith and works which could even lead Christians to worship well. No doubt non-Jewish slaves assisted in construction of the temple in Jerusalem and many non-Christians have worked in production studios creating contemporary Christian movies and worship albums. Is their contribution to the church any less significant? Thus, the question could be asked, would the lack of possession of the Holy Spirit have implications for the findings? Where Exodus demonstrates six qualities the technical artist must possess—the Spirit of God, skill, intelligence, knowledge, craftsmanship, and teaching ability—can any of these qualities—namely the Spirit—be lacking and still serve in building up the church? Secular artists can indeed move people to feel an "out-of-body experience" of sorts through excellent craftsmanship, without possessing the Holy Spirit themselves. Likewise, could the opposite be true? Could a non-believer be led to worship Christ through the creation of excellent art? Just as Christians view beautiful sunsets and natural wonders, declaring God's majesty behind creation, in the same way, many non-believers too gaze at the skies, recognizing there may something bigger than themselves at work. Most people, including those not of faith, believe just by experiencing the world around them—even a world void of obvious scriptural references—that there must be some god or master creator of the universe.[38] Is it possible, therefore, that the arts can have the same effect on a person? Is it possible that an individual can be so moved by a technical production of well-executed light, sound, and atmosphere—also void of obvious scriptural reference—that he or she cannot but proclaim that the experience is of God and in effect grow in communion with him? In other words, can and how can it be evangelical? Even though Scripture itself declares that general revelation in nature is indeed possible, a future study that focuses on the practical application of these principles played out may show that an artist as *imago Dei* can indeed create a way of meeting that leads believers and non-believers alike to recognize God's glory and handiwork through the arts.

---

[38] David Masci and Gregory A. Smith, "Is God Dead? No, but Belief Has Declined Slightly," *Pew Research*, April 7, 2016, http://www.pewresearch.org/fact-tank/2016/04/07/is-god-dead-no-but-belief-has-declined-slightly/. According to the 2016 Pew Research study, eighty-nine percent of Americans believe in God to some form or another, and sixty-three percent believe with absolute certainty.

Additionally, how might have these findings be impacted if other biblical passages were explored in the same way? Stated earlier, the one possible objection to these results is that they could be viewed as inconclusive due to the limited focus on only three specific biblical texts. How might expanding the study to other passages paint a wider portrait of the technical arts? While there are three passages explored here, I propose that these are not the only three that support these findings. How would other passages shape the outcome and offer other possible understandings for application in praxis?

This research began by springboarding off the current scholarly discussion in musical, liturgical worship studies. While the results are focused solely on technical artists, these findings could inform praxis for musicians, worship leaders, and other physical trades. During the writing of this book, it became apparent that this would likely equally apply to other craftspeople who serve the church, like masons, carpenters, architects, holiday decorators, and facilities teams. Likewise, how might these findings apply to Christians performing their craft outside the church—since the modern church is its people and not a building? Would painters, electricians, mechanics, web designers, and other technical trades be able to have their craft serve the church's mission should these findings be likewise applied? Furthermore, would or could the application outside the church serve a ministerial or evangelical purpose?

Last, this theology specifically focuses on the people who are tasked with the job of producing worship and purposefully ignored any specific tools of the trade. Future research would be well served to explore the actual technological equipment and methods that church techs use when performing their craft. How do the various output options affect the congregation's participation in the worship experience? For example, do purples make people connect better than reds or greens? Does a 6db difference in sound volume change how one "feels" the music? Does lowering the air-conditioning temperature better create a feeling of the Spirit at work? Do lower-third lyrics over live shots connect participants to the worship experience better than full-screen graphics? Are a couple lines of textual lyrics better than displaying the entire verse at once? Questions like these could be a natural next step in exploring applied technical arts.

These are just a few of the questions that future research can answer. Each of these could be a book in itself. Technical artists, worship leaders, pastors, and scholars alike, that is your challenge.

❧ 7 ❧

A FINAL WORD
TO CHURCH TECHNICAL ARTISTS

Since a biblically focused exegesis of the craft was severely lacking in technical arts circles prior to this study, I challenge that if technical artists themselves are not willing to stake their claim for theological validity within the church, how or why would other church leaders like pastoral staff, senior administration, and worship leaders do so? Technical artists must not only master their craft, but stake their claim as theologians themselves. It is their duty as God's contemporary, physical mediators between Christ and the church body. This book is the just first step in bridging the gap between theology and praxis. To church technical artists today: your work matters. Scripture both dictates it and demands it. You are today's Michelangelos telling God's story through your craft, moved by the Spirit of God, in response to your personal calling. With each audio patch, fader move, camera angle, lighting cue, projected image, and lyric slide advance, you are writing the next chapter in God's narrative for the church. You are producing worship.

SELECTED BIBLIOGRAPHY[1]

Ableman, Robert. "Without Divine Intervention: Contemporary Christian Music Radio and Audience Transference." *Journal of Media and Religion* 5, no. 4 (2006): 209–31.

Alexander, T. Desmond. "Authorship of the Pentateuch." In *Dictionary of the Old Testament: Pentateuch*, edited by T. Desmond Alexander and David W. Baker, 61–73. Downers Grove: InterVarsity Press, 2003.

-----. *Paradise to the Promised Land: An Introduction to the Pentateuch*. Grand Rapids: Baker Academic, 2012.

Allen, David L. *Hebrews*. NAC. Nashville: B&H Publishers, 2010.

Alluri, Vinoo, Petri Toiviainen, Iiro P. Jaaskelainen, Enrico Glerean, Mikko Sams, and Elvira Brattico. "Large-Scale Brain Networks Emerge from Dynamic Processing of Musical Timbre, Key and Rhythm." *NeuroImage* 59 (2012): 3677–89.

Anders, Max. *Holman New Testament Commentary: Galatians, Ephesians, Philippians, Colossians*. Nashville: B&H Publishers, 1999.

Anderson, Gary A. "Towards a Theology of the Tabernacle and Its Furniture." In *Text, Thought, and Practice in Qumran and Early Christianity*, edited by Ruth Clements and Daniel R. Schwartz, 159–94. Leiden: Brill Publishers, 2009.

Aniol, Scott. *Worship in Song: A Biblical Approach to Music and Worship*. Winona Lake: BMH Books, 2009.

-----. *Sound Worship: A Guide to Making Musical Choices in a Noisy World*. Fort Worth: Religious Affections Ministries, 2010.

-----. "Toward a Biblical Understanding of Culture." *The Artistic Theologian* 1 (2012): 40–56.

-----. "Worldview Bias and the Origin of Hebrew Worship." *Answers Research Journal* 8 (2015): 353–59.

Aranoff, Jeremy. "Torah: The Quintessential Blueprint: An Approach to Contemporary Jewish Architecture." Master's thesis, Carleton University, 2012.

Aristotle. *Rhetoric*. The Internet Classics Archive, translated by W.R. Roberts. Cambridge: Web Atomics, 1998.

Arnold, Bill T. (2003) "Pentateuchal Criticism, History of." In *Dictionary of the Old Testament: Pentateuch*, edited by T. Desmond Alexander and David W. Baker, 622–31. Downers Grove: InterVarsity Press, 2003.

Arnold, Clinton E. *Ephesians*. ZECNT. Grand Rapids: Zondervan, 2010.

Ashton, Mark and C. J. Davis. "Following in Cranmer's Footsteps" In *Worship by the Book*, edited by D. A. Carson, 64–135. Grand Rapids: Zondervan, 2002.

Attridge, Harold W. *The Epistle to the Hebrews: A Commentary on the Epistle to the Hebrews*. Philadelphia: Fortress Press, 1989.

-----. "The Psalms in Hebrews." In *The Psalms in the New Testament*, edited by Steve Moyise and Maarten J. J. Menken, 197–212. New York: T&T Clark, 2004.

---

[1] Over ten thousand individual media items related to the practice of church technical arts were surveyed during the documentary analysis research phase. These included: blog posts, trade articles, podcasts, social media posts, website content, industry newsletters, conference and trade show proceedings, and manufacturer training sessions. Items cited or referenced in this book are done so within the chapter footnotes, limiting the selected bibliography solely to that which is scholarly in nature.

-----. "God in Hebrews." In *The Epistle to the Hebrews and Christian Theology*, edited by Richard Bauckham. Grand Rapids: W. B. Eerdmans, 2009.

Attridge, Harold W. and Margot E. Fassler. *Psalms in Community: Jewish and Christian Textual, Liturgical, and Artistic Traditions*. Atlanta: Society of Biblical Literature, 2004.

Averbeck, Richard. E. "Worshiping God in Spirit." In *Authentic Worship: Hearing Scripture's Voice, Applying Its Truths*, edited by Herbert Bateman, 79–106. Grand Rapids, MI: Kregel Publications, 2002.

-----. "Tabernacle." In *Dictionary of the Old Testament: Pentateuch*, edited by T. Desmond Alexander and David W. Baker, 807–27. Downers Grove: InterVarsity Press, 2003.

Baker, David W. "Arts and Crafts." In *Dictionary of the Old Testament: Pentateuch*, edited by T. Desmond Alexander and David W. Baker, 49–53. Downers Grove: InterVarsity Press, 2003.

-----. "Source Criticism." In *Dictionary of the Old Testament: Pentateuch*, edited by T. Desmond Alexander and David W. Baker, 798–805. Downers Grove: InterVarsity Press, 2003.

Banjo, Omotayo O. and Williams, Kesha Morant. "A House Divided? Christian Music in Black and White." *Journal of Media and Religion* 10, no. 3 (2011): 115–37.

Bar-Efrat, Shimon. *Narrative Art in the Bible*. New York: T&T Clark, 2004.

Barnes, Albert, John Calvin, Adam Clarke, Matthew Henry, Charles H. Spurgeon, and John Wesley. *The Ultimate Commentary on Colossians: A Collective Wisdom on the Bible*. Seattle: Amazon Digital Services, 2016, Kindle.

Bauder, Kevin T. "Why Pastors Should Be Learned in Worship and Music." *The Artistic Theologian* 1 (2012): 3–15.

Bauckham, Richard. "The Divinity of Jesus Christ in the Epistle to the Hebrews." In *The Epistle to the Hebrews and Christian Theology*, edited by Richard Bauckham, Daniel R. Driver, Trevor A. Hart, and Nathan MacDonald, 15–36. Grand Rapids: Wm B. Eerdmans, 2009.

Bayne, Tim and Nagasawa, Yujin. "The Grounds of Worship." *Religious Studies* 42, no. 3 (2006): 299–313.

Beale, G. K. "Colossians." In *Commentary on the New Testament Use of the Old Testament*, edited by G. K. Beale and D. A. Carson, 841–70. Grand Rapids: Baker Academic, 2007.

-----. *We Become What We Worship: A Biblical Theology of Idolatry*. Downers Grove: IVP Academic, 2008.

-----. *Handbook on the New Testament use of the Old Testament: Exegesis and Interpretation*. Grand Rapids: Baker Academic, 2012.

Begbie, Jeremy. *Voicing Creation's Praise: Towards a Theology of the Arts*. Worcester: Billing & Sons Ltd., 1991.

-----. *Theology, Music, and Time*. Cambridge: Cambridge University Press, 2000.

-----. *Sounding the Depths: Theology Through the Arts*. London: SCM Press, 2002.

-----. *Resounding Truth: Christian Wisdom in the World of Music*. Grand Rapids: Baker Academic, 2007.

-----. "Faithful Feelings: Music and Emotion in Worship." In *Resonant Witness: Conversations between Music and Theology*, edited by Jeremy Begbie and Steven R. Guthrie, 323–54. Grand Rapids: Wm B. Eerdmans Publishing, 2011.

-----. "The Holy Spirit at Work in the Arts: Learning from George Herbert." *Interpretation: A Journal of Bible and Theology* 66, no. 1 (2012): 41–54.

Begbie, Jeremy and Steven R. Guthrie. (2011) "Introduction." In *Resonant Witness: Conversations between Music and Theology*, edited by Jeremy Begbie and Steven R. Guthrie, 1–26. Grand Rapids: Wm B. Eerdmans Publishing, 2011.

Benjamin, Walter. "The Work of Art in the Age of Mechanical Reproduction." In *Illuminations*, edited by Hannah Arendt, translated by Harry Zohn, 1–26. New York: Schocken Books, 1969.

Bentley, Joshua. "A Uses and Gratification Study of Contemporary Christian Radio Web Sites." *Journal of Radio & Audio Media* 19, no. 1 (2012): 2–16.

Bergeson, Kevin D. "Sanctuary as Cinema? Screens Should Not Block the Story." *Word & World* 32, no. 3 (2012): 303–4.

Bezuidenhoudt, Jacobus. "The Renewal of Reformed Worship through Retrieving the Tradition and Ecumenical Openness." Master's thesis, University of Cape Town, 2000.

Blackmon, Johnathan. "Scripture, Shekinah, and Sacred Song: How God's Word and God's Presence Should Shape the Song of God's People." *Artistic Theologian* 1 (2012): 25–39.

Bloch, Lee J. *Worship from Backstage: A Biblical Perspective: A Guide for Technical Arts Ministry & Facility Planning*. USA: Lee J. Bloch, 2008, Kindle.

Blum, Gerd. "Vasari on the Jews: Christian Canon, Conversion, and the Moses of Michelangelo." *The Art Bulletin* 95, no.4 (2013): 557–77.

Bonhoeffer, Dietrich. *Letters and Papers from Prison*. New York: Touchstone, 1997.

Booth, Eric. "Seeking Definition: What is a Teaching Artist?" *Teaching Artist Journal* 1, no. 1 (2003): 5–12.

Borthwick, Alastair, Trevor Hart, and Anthony Monti. "Musical Time and Eschatology." In *Resonant Witness: Conversations between Music and Theology*, edited by Jeremy Begbie and Steven R. Guthrie, 271–94. Grand Rapids: Wm B. Eerdmans Publishing, 2011.

Bradshaw, Paul F. *The Search for the Origins of Christian Worship: Sources and Methods for the Study of Early Liturgy*. New York: Oxford University Press, 2002.

-----. *Early Christian Worship: A Basic Introduction to Ideas and Practice*. Collegeville: Liturgical Press, 2010.

Brewster, Karen and Melissa Shafer. *Fundamentals of Theatrical Design*. New York: Allworth Press, 2011.

Bromiley, Geoffrey W. "Evangelicals and Theological Creativity." *Themelios* 5, no. 1 (1979): 4–8.

Brown, Campbell and Yujin Nagasawa. "I Can't Make You Worship Me." *Ratio* 20 (2005): 236–40.

Brown, Frank Burch. *Religious Aesthetics: A Theological Study of Making and Meaning*. Princeton: Princeton University Press, 1989.

-----. *Good Taste, Bad Taste, and Christian Taste: Aesthetics in Religious Life*. New York: Oxford University Press, 2000.

-----. "Music: Religious Music in the West." *Encyclopedia of Religion*, edited by Lindsay Jones, 6307–14. Detroit: Macmillan Reference USA, 2005.

-----. *Inclusive Yet Discerning: Navigating Worship Artfully*. Grand Rapids, MI: Eerdmans, 2009.

-----. "Introduction: Mapping the Terrain of Religion and the Arts." In *The Oxford Handbook of Religion and the Arts*, edited by Frank Burch Brown, 1–21. New York: Oxford University Press, 2014.

-----. "Musical Ways of Being Religious." In *The Oxford Handbook of Religion and the Arts*, edited by Frank Burch Brown, 109–29. New York: Oxford University Press, 2014.

Bruce, F. F. *The Epistles to the Colossians, to Philemon, and to the Ephesians*. Grand Rapids: Wm. B. Eerdmans, 1984.

-----. "The Son of Man the Savior and High Priest of His People 2:10–18." *The Epistle to the Hebrews*, edited by Gordon D. Fee. Grand Rapids: W. B. Eerdmans, 1990.

Brueggemann, Dale A. "Psalms 4: Titles." In *Dictionary of the Old Testament: Wisdom, Poetry & Writings*, edited by Tremper Longman III and Peter Enns, 613–21. Downers Grove: IVP

Academic, 2008.

Buckley, Paul. "Psalms." In *It Was Good: Making Music to the Glory of God*, edited by Ned Bustard, 5593–915. Baltimore: Square Halo Books, 2013.

Bullock, C. Hassell. *An Introduction to the Old Testament Poetic Books*. Chicago: Moody Press, 1998.

Burgeson, K. D. "Sanctuary as Cinema? Screens Should Not Block the Story." *Word & World* 32, no. 3 (2012): 303–5.

Burreson, Kent J. "Beyond Style: The Worship of Christ's Body within Cultural Diversity." *Cross Accent* 21, no. 3 (2013): 8–17.

Cain, Andrew and Angharad Parry Jones. "Flat Screens and Rood Screens: The Integration of Audio-Visual Technology into Traditional Worship." *Modern Believing* 50, no. 2 (2009): 40–42.

Calvin, John. *Preface to the Psalter*. Translated by Kassel Baeroenreiter-Verlag. Geneva: La Société des Concerts de la Cathédrale de Lausanne, 1935.

-----. *Commentary on Colossians*. Waikato: Titus Books, 2012.

Calvin, John and Charles Bingham. *Commentaries on the Four Last Books of Moses Arranged in the Form of a Harmony*. Bellingham: Logos Bible Software, 2010.

Calvin, John and John Owen. *Commentary on the Epistle of Paul the Apostle to the Hebrews*. Bellingham: Logos Bible Software, 2010.

Campbell, Drew. *Technical Theater for Nontechnical People*. New York: Allworth, 2004.

Campbell, Iain D. *Opening Up Exodus*. Opening Up Commentary. Leominster: Day One Publications, 2006.

Carson, D. A. *Worship: Adoration and Action*. Grand Rapids: Baker Book House, 1993.

-----. *Worship by the Book*. Grand Rapids: Zondervan, 2002.

-----. *Worship: Adoration and Action*. Eugene: Wipf and Stock Publishers, 2002.

Carson, D. A. and Douglas J. Moo. *An Introduction to the New Testament*. Grand Rapids: Zondervan, 2005.

Carson, D. A., Douglass J. Moo, and Andrew David Naselli. *Introducing the New Testament: A Short Guide to Its History and Message*. Grand Rapids: Zondervan, 2010.

Carter, Paul. *Backstage Handbook: An Illustrated Almanac of Technical Information*. New York: Broadway Press, 1998.

Celaya, Chaz. *The Sound Guide: A Resource for Audio Non-Professionals*. Costa Mesa, CA: Positive Infinity Publishing, 2013.

Chapell, Bryan. *Christ-Centered Worship: Letting the Gospel Shape Our Practice*. Grand Rapids: Baker Academic, 2009.

Charlesworth, James H. *The Earliest Christian Hymnbook: The Odes of Solomon*. Eugene: Cascade Books, 2009.

Charney, Davida. "Maintaining Innocence Before a Divine Hearer: Deliberative Rhetoric in Psalm 22, Psalm 17, and Psalm 7." *Biblical Interpretation* 21, no. 1 (2013): 33–63.

Cheetham, David. "The Bible and Music." *Scripture Bulletin* 28, no. 2 (1998): 59–68.

-----. *Ways of Meeting and the Theology of Religions*. Burlington: Ashgate Publishing Company, 2013.

Chester, Andrew. "High Christology – Whence, When and Why?" *Early Christianity* 2, no.1 (2011): 22–50.

Cheung, Vincent. *Commentary on Colossians*. VincentCheung.com, 2008.

Chua, Daniel K. L. "Music as the Mouthpiece of Theology." In *Resonant Witness: Conversations between Music and Theology*, edited by Jeremy Begbie and Steven R. Guthrie, 137–61. Grand Rapids: Wm B. Eerdmans Publishing, 2011.

Clowney, Edmund. "Living Art: Christian Experience and the Arts." In *God and Culture*, edited by D. A. Carson and John D. Woodbridge. Grand Rapids: Wm. B. Eerdmans Publishing Company, 1993.

Cobb, Kelton. *The Blackwell Guide to Theology and Popular Culture*. Malden: Blackwell Publishing, 2005.

Cockerill, Gareth L. *The Epistle to the Hebrews*. NICNT. Grand Rapids: Eerdmans, 2012.

Coffin, Henry Sloane. "Worship." *Christian Education* 12, no. 4 (1929): 218–20.

Cosper, Mike. *Rhythms of Grace: How the Church's Worship Tells the Story of the Gospel*. Wheaton: Crossway, 2013.

Costa, Tony. *Worship and the Risen Jesus in the Pauline Letters*. Studies in Biblical Literature. New York: Peter Lang AG, 2013.

Crabtree, Jeff. "Colossians 3:16 and the Deliberate Growth of Disciples." DMin dissertation, Acadia University, 2005.

Cray, Wesley D. "Omniscience and Worthiness of Worship." *International Journal for Philosophy of Religion* 70, no. 2 (2011): 147–53.

Crouch, Andy. "The Gospel: How Art Is a Gift, a Calling, and an Obedience?" In *For Beauty of the Church*, edited by W. David O Taylor. Grand Rapids: Baker Books, 2010.

Dawn, Marva J. *Reaching Out Without Dumbing Down: A Theology of Worship for the Turn-of-the-Century Culture*. Grand Rapids: W.B. Eerdmans, 1995.

-----. *A Royal Waste of Time: The Splendor of Worshiping God and Being Church for the World*. Grand Rapids: W.B. Eerdmans, 1999.

Dawn, Marva J. and Daniel Taylor. *How Shall We Worship?: Biblical Guidelines for the Worship Wars*. Wheaton: Tyndale House Publishers, 2003.

De Gruchy, John W. "Art, Morality, and Justice." In *The Oxford Handbook of Religion and the Arts*, edited by Frank Burch Brown, 418–32. New York: Oxford University Press, 2014.

Deneault, Wayne. *Technology for God's Kingdom: An Evaluation*. Canadian Centre for Worship Studies. CCWS, 2003.

deSilva, David A. "Responding to God's Word and Work in the Son: 1:1–2:18." In *Perseverance in Gratitude: A Socio-Rhetorical Commentary on the Epistle "to the Hebrews,"* 83–130. Grand Rapids: W.B. Eerdmans, 2000.

Detwiler, David F. "Church Music and Colossians 3:16." *Bibliotheca Sacra* 158 (July–September 2001): 347–69.

Dillenberger, John. "Contemporary Theologians and the Visual Arts." *Journal of the American Academy of Religion* 53, no. 4 (1985): 599–615.

Docherty, Susan E. *The Use of the Old Testament in Hebrews: A Case Study in Early Jewish Bible Interpretation*. Tübingen: Mohr Siebeck, 2009.

Dozeman, Thomas B. *Exodus*. ECC. Grand Rapids: Eerdmans, 2009.

Dunn, James D. G. *Did the First Christians Worship Jesus?: The New Testament Evidence*. Louisville: Westminster John Knox Press, 2010.

Durham, John I. *Exodus*. Word Biblical Commentary. Waco: Word, Incorporated, 1987.

Dyson, Freeman. "The Twenty-First Century." In *Infinite in All Directions: Gifford Lectures Given at Aberdeen, Scotland, April-November 1985*, edited by Freeman Dyson, 270–74. New York: Harper & Row, 2004.

Ehorn, Seth M. "Colossians, Letter to the, Critical Issues." In *The Lexham Bible Dictionary*, edited by John D. Barry. Bellingham: Lexham Press, 2016.

Ellingworth, Paul. *The Epistle to the Hebrews: A Commentary on the Greek Text*. Grand Rapids: W.B. Eerdmans, 1993.

Enns, Peter. "The Spirit's Ministry in the Old Testament." In *The Moody Handbook of Theology*, 260–62. Chicago: Moody Press, 1989.

Erlenbach, Bill. *Reflecting Truth in Beauty.* Canadian Centre for Worship Studies. CCWS, 2007.

Ermer, Gayle. "Responsible Engineering and Technology." In *Delight in Creation: Scientists Share Their Work with the Church,* edited by Deborah Haarsma and Scott Hoezee, 127–39. Grand Rapids: Calvin Theological Seminary, 2012.

Evans, Mark. *Open Up the Doors: Music in the Modern Church.* London: Equinox Publishing Ltd, 2006.

Farley, Michael A. "What is 'Biblical' Worship? Biblical Hermeneutics and Evangelical Theologies of Worship." *Journal of the Evangelical Theological Society* 51, no. 3 (2008): 591–613.

Farley, Todd. "Theater in Liturgy as Actio Divina—God's Self-Performance." *Liturgy* 24, no. 1 (2008): 33–39.

Fassler, Margot E., Jane Huber, and Sachin Ramabhadran, S. *Joyful Noise: Psalms in Community.* Atlanta: Society of Biblical Literature, 2008.

Fee, Gordon D. *God's Empowering Presence: The Holy Spirit in the Letters of Paul.* Peabody: Hendrickson Publishers, 1994.

Ferebee, Susan, James Davis, Martine Bates Shart, and Stephen Beyer. "The Relationship of Bible Medium to Bible Perception." *The Exchange* 4, no. 2 (2015): 86–98.

Ferretter, Luke. "The Power and the Glory: The Aesthetics of the Hebrew Bible." *Literature & Theology* 18, no. 2 (2004): 123–38.

Fesko, J. V. *Christ and the Desert Tabernacle.* Darlinghton: EP Books: 2012.

Fine, Steven. *Art, History, and the Historiography of Judaism in Roman Antiquity.* Boston: Brill, 2014.

Foley, Edward. *Foundations of Christian Music: The Music of Pre-Constantinian Christianity.* Collegeville: Liturgical Press, 1996.

Forest, Tom. "Proper Worship." *Tabletalk Magazine* (July 1996): 45.

Fowl, Stephen E. *Ephesians: A Commentary.* Louisville: Westminster John Knox Press, 2012.

Frame, John. *Worship in Spirit and Truth.* Phillipsburg: P&R Publishing, 1996.

Fretheim, Terence E. "Because the Whole Earth is Mine: Theme and Narrative in Exodus." *Interpretation* 50, no. 3 (1996): 229–39.

Gaiser, Frederick J. "Keeping It Real." *Word & World* 32, no. 3 (2012): 215–16.

Ganga-Persad, Deve. *A Theology of Image (For Use of Projection Technology in Congregational Settings).* Canadian Centre for Worship Studies. CCWS, 2007.

Garrett, Duane. *A Commentary on Exodus.* Kregel Exegetical Library. Grand Rapids: Kregel Academic, 2013.

Gasque, Laurel. "The Christian Stake in the Arts: Toward a Missiology of Western Culture." *Evangelical Review of Theology* 24, no. 3 (2000).

Gelardini, Gabriella. "Hebrews, Homiletics, and Liturgical Scripture Interpretation." In *Reading the Epistle to Hebrews: A Resource for Students,* edited by Eric Mason and Kevin McCruden, 121–44. Atlanta: Society of Biblical Literature, 2011.

George, Mark K. *Israel's Tabernacle as Social Space.* Atlanta: SBL, 2009.

-----. "Analyzing Embodied Space in Ancient Israel's Tabernacle: A Biblical Studies Perspective." In *Religious Representation in Place: Exploring Meaningful Spaces at the Intersection of the Humanities and Sciences,* edited by Mark K. George and Daria Pezzoli-Olgiati, 63–74. New York: Palgrave Macmillan, 2014.

Gieck, Lorn. *Music's Role in the Revelation Response Dialogue of Christian Worship.* Canadian Centre for Worship Studies. CCWS, 2008.

Gillette, J. Michael. *Theatrical Design and Production: An Introduction to Scene Design and Construction, Lighting, Sound, Costume, and Makeup.* New York: McGraw-Hill, 2013.

Glodo, Michael. "Singing with the Savior." *Reformed Quarterly* 17, no. 1 (1998): 1–5.

Gormly, Eric. "Evangelizing Through Appropriation: Toward a Cultural Theory on the

Growth of Contemporary Christian Music." *Journal of Media and Religion* 2, no. 4 (2003): 251–65.

Grimes, Brad. *CTS Certified Technology Specialist Exam Guide*. New York: McGraw-Hill Education, 2013.

Grintz, Yehoshua. "Bezalel." In *Encyclopedia Judaica*, edited by Fred Skolnik and Michael Berenbaum, 557. Farmington Hills: Keter Publishing House, Ltd., 2007.

Grit, Betty. "Welcoming the Arts in Worship." 2013 Calvin Symposium on Worship. Grand Rapids: Calvin College, 2013.

Gromacki, Robert Glenn. "The Exaltation of Man: Hebrews 2:1–18." In *Stand Bold in Grace: An Exposition of Hebrews*, 37–53. Grand Rapids: Baker Book House, 1984.

Grudem, Wayne. *Systematic Theology: An Introduction to Biblical Doctrine*. Grand Rapids: Zondervan, 2004.

Guthrie, George H. "Hebrews." In *Commentary on the New Testament Use of the Old Testament*, edited by G. K. Beale and D. A. Carson, 919–96. Grand Rapids: Baker Academic, 2007.

Guthrie, Steven R. "Temples of the Spirit: Worship as Embodied Performance." In *Faithful Performances: The Enactment of Christian Identity in Theology and the Arts*, edited by Trevor Hart and Steven R. Guthrie, 91–107. Burlington: Ashgate Publishing Group, 2007.

-----. *Creator Spirit: The Holy Spirit and the Art of Becoming Human*. Grand Rapids: Baker Academic, 2011.

-----. "The Wisdom of Song." In *Resonant Witness: Conversations between Music and Theology*, edited by Jeremy Begbie and Steven R. Guthrie, 382–407. Grand Rapids: Wm B. Eerdmans Publishing, 2011.

-----. "Love the Lord with All Your Voice: Singing Is a Forgotten—But Essential—Spiritual Discipline." *Christianity Today* 57 (2013): 44–47.

Hagner, Donald Alfred. "The Full Humanity of the Son of God." In *Encountering the Book of Hebrews: An Exposition*, 53–62. Grand Rapids: Baker Academic, 2002.

Hamilton, Victor P. *Handbook on the Historical Books*. Grand Rapids: Baker Academic, 2001.

-----. *Exodus: An Exegetical Commentary*. Grand Rapids: Baker Academic, 2011.

Harris, Clement Antrobus. "On the Divine Origin of Musical Instruments in Myths and Scriptures." *The Musical Quarterly* 8, no. 1 (1922): 69–75.

Harris, Dale. *The Curious Works of Bezalel: Reconsidering the Role of the Artist in the Emergent Church*. Canadian Centre for Worship Studies. CCWS, 2006.

-----. *The Things of God to Man and the Things of Man to God: Worship and the Mediation of Christ*. Canadian Centre for Worship Studies. CCWS, 2007.

Hart, Trevor. "Art, Performance and the Practice of Christian Faith." In *Faithful Performances: The Enactment of Christian Identity in Theology and the Arts*, edited by Trevor Hart and Steven R. Guthrie, 1–9. Burlington: Ashgate Publishing Group, 2007.

-----. "The Sense of an Ending: Finitude and the Authentic Performance of Life." In *Faithful Performances: The Enactment of Christian Identity in Theology and the Arts*, edited by Trevor Hart and Steven R. Guthrie, 167–86. Burlington: Ashgate Publishing Group, 2007.

Hartley, J. E. "Holy and Holiness, Clean and Unclean." In *Dictionary of the Old Testament: Pentateuch*, edited by T. Desmond Alexander and David W. Baker, 420–31. Downers Grove: InterVarsity Press, 2003.

Hawthorne, Steven C. "Let All the Peoples Praise Him: Toward a Teleological Paradigm of the Missio Dei." Ph.D. dissertation, Fuller Theological Seminary, 2013.

Heaney, Maeve Louise. *Music as Theology: What Music Says about the Word*. Eugene: Pickwick Publications, 2012.

Hearon, Holly E. "Music as a Medium of Oral Transmission in Jesus Communities." *Biblical*

Theology Bulletin 43, no. 4 (2013): 180–90.

Heil, John Paul. *Letters of Paul as Rituals of Worship*. Eugene: Cascade Books, 2011.

-----. *Worship in the Letter to the Hebrews*. Eugene: Cascade Books, 2011.

Heinemann, Mark H. "An Exposition of Psalm 22." *Bibliotheca Sacra* 147 (July 1990): 286–308.

Hendrix, Ralph E. "A Literary Structural Overview of Exod 25–40." *Andrews University Seminary Studies* 30, no. 2 (1992): 123–38.

-----. "The Use of Miskan and 'Ohel Mo'ed in Exodus 25–40." *Andrews University Seminary Studies* 30, no. 1 (1992): 3–13.

Herring, Brad. *Sound, Lighting and Video: A Resource for Worship*. San Diego: Focal Press, 2009.

Hess, Richard. "Bezalel and Oholiab: Spirit and Creativity." In *Presence, Power and Promise: The Role of the Spirit of God in the Old Testament*, edited by David G. Firth and Paul D. Wegner, 161–74. Downers Grove: IVP Academic, 2011.

Hiebert, Theodore. "Altars of Stone and Bronze: Two Biblical Views of Technology." *Mission Studies* XV-2, no. 30 (1998): 75–84.

Hillman, Jordan Jay. "Exodus." In *The Torah and Its God: A Humanist Inquiry*, 149–247. Amherst: Prometheus Books, 2001.

Hoehner, Harold W. *Ephesians: An Exegetical Commentary*. Grand Rapids: Baker Academic, 2002.

Holmes, Michael W. *The Greek New Testament: SBL Edition*. Bellingham: Lexham Press, 2013.

Homan, Michael M. *To Your Tents, O Israel!: The Terminology, Function, Form, and Symbolism of Tents in the Hebrew Bible and the Ancient Near East*. Boston: Brill, 2002.

-----. "The Tabernacle and the Temple in Ancient Israel." *Religion Compass* 1, no. 1 (2007): 38–49.

Hughes, R. Kent. *Colossians and Philemon: The Supremacy of Christ*. Westchester: Crossway Books, 1989.

-----. *Ephesians: The Mystery of the Body of Christ*. Wheaton: Crossway Books, 1990.

-----. "Free Church Worship: The Challenge of Freedom." In *Worship by the Book*, edited by D. A. Carson, 136–92. Grand Rapids: Zondervan, 2002.

Hurowitz, Victor. "The Priestly Account of Building the Tabernacle." *American Oriental Society* 105, no. 1 (1985): 21–30.

Hurtado, Larry W. "The Origins of the Worship of Christ." *Themelios* 19, no. 2 (1994): 4–8.

-----. "The Binitarian Shape of Early Christian Worship." In *The Jewish Roots of Christological Monotheism: Papers from the St. Andrews Conference on the Historical Origins of the Worship of Jesus*, edited by Carey C. Newman, James R. Davila, and Gladys S. Lewis, 187–213. Boston: Brill, 1999.

-----. "Worship, NT Christian," In *The New Interpreter's Dictionary of the Bible*, edited by K. D. Sakenfeld, 910–23. Nashville: Abingdon Press.

Huyser-Honig, Joan. "Technology in Worship: Beyond PowerPoint." Calvin Institute of Christian Worship. Grand Rapids: Calvin College, 2004.

-----. "Contemporary Worship Music Matters." Calvin Institute of Christian Worship. Grand Rapids: Calvin College, 2005.

-----. "Digital Storytelling: Use Multimedia in Worship to Enhance, Not Replace." Calvin Institute of Christian Worship. Grand Rapids: Calvin College, 2007.

Illiff, M. "Worship Wars within Reformed Evangelicalism." 2004 Evangelical Identities Conference. London: Kings College, 2004.

Ito, John Paul. "On Music, Mathematics, and Theology." In *Resonant Witness: Conversations between Music and Theology*, edited by Jeremy Begbie and Steven R. Guthrie, 109–34. Grand Rapids: Wm B. Eerdmans Publishing, 2011.

Iwand, Hans J. "The Appropriation of New Righteousness." In *The Righteousness of Faith according to Luther*, edited by Virgil F. Thompson, 76–79. Eugene: Wipf & Stock, 2008.

Janzen, James D. *Psalms, Hymns and Spiritual Songs: The Road to Unity and Spiritual Maturity*. Victoria: Friesen Press, 2015, Kindle.

Janzen, Waldemar. *Exodus*. Scottdale: Herald, 2000.

Jobes, Karen H. "The Use of the Old Testament in Hebrews." In *Letters to the Church: A Survey of Hebrews and the General Epistles*. Grand Rapids: Zondervan, 2011.

Johnson, Birgitta. "Back to the Heart of Worship: Praise and Worship Music in a Los Angeles African-American Megachurch." *Black Music Research Journal* 31, no. 1 (2011): 105–29.

Johnstone, William. *Exodus 20–40*. Smyth & Helwys Bible Commentary. Macon: Smyth & Helwys, 2014.

Jones, Paul S. *What is Worship Music? Basics of the Faith Series*. Phillipsburg: P&R Publishing, 2010.

Joslin, Barry. "Raising the Worship Standard: The Translation and Meaning of Colossians 3:16 and Implications for Our Worship." *Southern Baptist Journal of Theology* 17, no. 3 (2013): 50–59.

Kapelrud, Arvid S. "Temple Building, a Task for Gods and Kings." *Orientalia* 32 (1963): 56–62.

Kauflin, Bob. *Worship Matters*. Wheaton: Crossway Books, 2008.

Kearney, Peter J. "Creation and Liturgy: The P Redaction of Exodus 25–40." *Zeitschrift für die Alttestamentliche Wissenschaft* 89 (1997): 375–87.

Keith, Shannon D. "The Perceived Role of Music in the Pentecostal and Charismatic Worship Experience." Master's thesis, University of Louisiana and Lafayette, 2013.

Keller, Tim J. "What It Takes to Worship Well." *LEADERSHIP Journal* 15, no. 4 (1994).

-----. "Worship Worthy of the Name." *Changing Lives through Preaching and Worship: 30 Strategies for Powerful Communication*, edited by Marshall Shelley, 178–185. Nashville: Random House, 1995.

-----. "Reformed Worship in the Global City." In *Worship by the Book*, edited by D. A. Carson, 193–250. Grand Rapids: Zondervan, 2002.

-----. "Why We Need Artists." In *It Was Good: Making Art to the Glory of God*, edited by Ned Bustard. Baltimore: Square Halo Books, 2007, Digital edition.

Kelley, B. "The Use of Technology in the Global Church." 2008 Christian Engineering Educators Conference. Beaver Falls: Geneva College, 2008.

Kenneson, Philip. "Gathering: Worship, Imagination, and Formation." In *The Blackwell Companion to Christian Ethics*. Stanley Hauerwas and Samuel Wells, 53–67, Malden, MA: 2004.

Kidd, Reggie M. *With One Voice: Discovering Christ's Song in Our Worship*. Grand Rapids: Baker Books, 2005.

Kidwell, Jeremy. *The Theology of Craft and the Craft of Work: From Tabernacle to Eucharist*. New York: Routledge, 2016.

Kitchen, Kenneth A. "Egypt's Impact on Early Hebrew Community and Culture." In *Dictionary of the Old Testament: Pentateuch*, edited by T. Desmond Alexander and David W. Baker, 212–14. Downers Grove: InterVarsity Press, 2003.

Klein, Ralph. "Back to the Future: The Tabernacle in the Book of Exodus." *Interpretation* 50, no. 3 (1996): 264–76.

Koester, Craig R. *Hebrews: A New Translation with Introduction and Commentary*. New York: Doubleday, 2001.

Köstenberger, Andreas J. and Richard Duane Patterson. *Invitation to Biblical Interpretation: Exploring the Hermeneutical Triad of History, Literature, and Theology*. Grand Rapids: Kregel

Publications, 2011.

Koster, Steven J. "Visual Media Technology in Christian Worship." Master's thesis, Michigan State University, 2003.

Kramer, Ben. *The Revelation/Response Dialogue in Christian Worship*. Canadian Center for Worship Studies. CCWS, 2006.

Kumorek, Jim. "Case Study: The Pastor's Guide to Lighting." *WorshipFacilities.com*, EH Publishing, Inc., 2006.

Lane, William L. *Hebrews 1–8*. WBC. Dallas: Word Books, 1991.

-----. "Hebrews." In *Dictionary of the Later New Testament and Its Developments*, edited by Ralph P. Martin and Peter H. Davids, 443–58. Downers Grove: InterVarsity Press, 1997.

Lange, John Peter, Philip Schaff, Karl Braune, and M. B. Riddle. *A Commentary on the Holy Scriptures: Colossians*. Bellingham: Logos Bible Software, 2008.

-----. *A Commentary on the Holy Scriptures: Ephesians*. Bellingham: Logos Bible Software, 2008.

Lange, John Peter, Philip Schaff, Charles M. Mead. *A Commentary on the Holy Scriptures: Exodus*. Bellingham: Logos Bible Software, 2008.

Lange, John Peter, Philip Schaff, Bernhard Moll, and A. C. Kendrick. *A Commentary on the Holy Scriptures: Hebrews*. Bellingham: Logos Bible Software, 2008.

Lea, Thomas D. "Hebrews." In *Holman Concise Bible Commentary*, edited by D. S. Dockery. Nashville: B&H Publishers, 1998.

-----. *Hebrews & James*. HNTC. Nashville: B&H Publishers, 1999.

Leder, Arie. "Reading Exodus to Learn and Learning to Read Exodus." *Calvin Theological Journal* 34 (1999): 11–35.

Leithart, Peter J. "Making and Mis-Making: Poiesis in Exodus 25–40." *International Journal of Systematic* Theology 2, no. 3 (2000): 307–18.

Levine, Baruch A. 'The Descriptive Tabernacle Texts of the Pentateuch." *Journal of the American Oriental Society* 85, no. 3 (1965): 307–18.

Levison, John R. *Filled with the Spirit*. Grand Rapids: W. B. Eerdmans, 2009.

Lindenbaum, John. "The Pastoral Role of Contemporary Christian Music: The Spiritualization of Everyday Life in a Suburban Evangelical Megachurch." *Social & Cultural Geography* 13, no. 1 (2012): 69–88.

Lindenbaum, John. 'The Neoliberalization of Contemporary Christian Music's New Social Gospel." *Geoforum* 44, no. 1 (2013): 112–19.

Lindsay, T. S. "Singing as Part of the Worship of the Early Church." *The Irish Church Quarterly* 4, no. 15 (1911): 244–54.

Lindsley, Art. "Gifts of the Spirit in the Old Testament." *Institute for Faith, Work, & Economics* (2013): 1–6.

Longman III, Tremper. "Spirit and Wisdom." In *Presence, Power and Promise: The Role of the Spirit of God in the Old Testament*, edited by David G. Firth and Paul D. Wegner, 95–110. Downers Grove: IVP Academic, 2011.

Lucarini, Dan. *Why I Left the Contemporary Christian Music Movement: Confessions of a Former Worship Leader*. Webster: Evangelical Press, 2002.

-----. *It's Not about the Music: A Journey into Worship*. Carlisle: EP Books, 2010.

Lundin, Roger. "Doubt and Belief in Literature." In *The Oxford Handbook of Religion and the Arts*, edited by Frank Burch Brown, 433–49. New York: Oxford University Press, 2014.

Lynch, Gordon. *Understanding Theology and Popular Culture*. Oxford: Blackwell Publishing Ltd., 2005.

MacArthur, John. "With Hearts and Minds and Voices." *Christian Research Journal* 23, no. 2 (2001): 1–7.

-----. *Worship: The Ultimate Priority*. Chicago: Moody Publishers, 2012.

Mackie, Scott D. "Heavenly Sanctuary Mysticism in the Epistle to the Hebrews." *The Journal of Theological Studies* 62, no. 1 (2011): 77–117.

Makujina, John. "Forging Musical Boundaries: The Contribution of 1 Corinthians 14:6–11 and Exodus 32:17–18 to a Christian Philosophy of Music." *The Artistic Theologian* 2 (2013): 1–63.

Man, Ron. "Another Helper: The Holy Spirit and Worship." *Worship Notes* 1, no. 9 (2006): 1–3.

-----. "Jesus Our Worship Leader: Proclamation and Praise." *Worship Notes* 1, no. 8 (2006): 1–2.

-----. *Proclamation and Praise: Hebrews 2:12 and the Christology of Worship*. Eugene: Wipf & Stock, 2007.

-----. "Biblical Principles of Worship & Their Application to Local Church Ministry." *Worr.org*. Worship Resources, 2009.

-----. "Jesus, Our True Worship Leader." *The Artistic Theologian* 2 (2013): 4–16.

-----. "Revelation and Response: The Pattern of True Worship." *The Evangelical Theological Society*, Atlanta: November 18, 2015.

Manser, Martin H. *Dictionary of Bible Themes: The Accessible and Comprehensive Tool for Topical Studies*. London: Martin Manser, 2009.

Marple, Norman C. "Worship that Speaks to the Core of Our Being." DMin dissertation, St. Stephen's College, 2008.

Martin, Ralph P. "Worship." In *Dictionary of Paul and His Letters*, edited by Gerald F. Hawthorne, Ralph P. Martin, and Daniel G. Reid, 982–90. Downers Grove: InterVarsity Press, 1993.

Martin, Ryan. "More on the Hymnal vs. Screen Thing." *Religious Affections Ministries*, February 14, 2012, http://religiousaffections.org/articles/articles-on-culture/more-on-the-hymnal-vs-screen-thing/.

Masci, David and Gregory A. Smith. (2016) "Is God Dead? No, but Belief Has Declined Slightly." *PewResearch.org*, April 7, 2016, http://www.pewresearch.org/fact-tank/2016/04/07/is-god-dead-no-but-belief-has-declined-slightly/.

McCrindle Research. *Generations Defined*. New South Wales: UNSW Press, 2011.

McKinley, John E. "Psalms 16, 22, and 110. Historically Interpreted as Referring to Jesus." *Perichoresis* 10, no. 2 (2012): 207–21.

McKnight, Edgar V. and Christopher Lee Church. "The Humiliation and Exaltation of Jesus," In *Hebrews-James*, 65–80. Macon: Smyth & Helwys, 2004.

McNaughton, Ian S. *Opening Up Colossians and Philemon*. Leominster: Day One Publications, 2006.

McNiff, Jean and Jack Whitehead. *You and Your Action Research Project*. New York: Routledge, 2010.

Mehta, Ravi, Rui Juliet Zhu, and Amar Cheema. "Is Noise Always Bad? Exploring the Effects of Ambient Noise on Creative Cognition." *Journal of Consumer Research* 39, no. 4 (2012): 784–99.

Melick, Richard R. *Philippians, Colossians, Philemon*. NAC. Nashville: B&H Publishers, 1991.

Merrill, E. H. "Image of God." In *Dictionary of the Old Testament: Pentateuch*, edited by T. Desmond Alexander and David W. Baker, 441–45. Downers Grove: InterVarsity Press, 2003.

Messenger, William. "Vocational Overview." Theology of Work Project, 2010

-----. "Calling in the Theology of Work." *Journal of Markets & Morality* 14, no. 1 (2011): 171–87.

Meyers, Carol. *Exodus*. The New Cambridge Bible Commentary. New York: Cambridge University Press, 2005.

Michaels, J. R. "Commentary on Hebrews." In *Cornerstone Bible Commentary: 1 Timothy, 2 Timothy, Titus, and Hebrews*, edited by P. W. Comfort. Carol Steam: Tyndale House Publishers, 2009.

Middleton, J. Richard. *The Liberating Image: The Imago Dei in Genesis 1*. Grand Rapids: Brazos Press, 2005.

Miller, Steve. *The Contemporary Christian Music Debate: Worldly Compromise or Agent of Renewal?* Acworth: Wisdom Creek Press, 2011.

Miller, Stephen. *Worship Leaders: We Are Not Rock Stars*. Chicago: Moody Publishers, 2013.

Mitchell, Alan C. "A Merciful and Faithful High Priest." In *Hebrews*, edited by Daniel J. Harrington. Collegeville: Liturgical Press, 2007.

Moltmann, Jürgen. *God in Creation: A New Theology of Creation and the Spirit of God*. San Francisco: Harper & Row, 1991.

Moo, Douglas J. *The Letters to the Colossians and to Philemon*. The Pillar New Testament Commentary. Grand Rapids: William B. Eerdmans Pub. Co., 2008.

Morschauser, Scott N. "Created in the Image of God: The Ancient Near Eastern Background of the Imago Dei." *Theology Matters* 3, no. 6 (1997): 1–9.

Mullins, Phil. "Imagining the Bible in Electronic Culture." *Religion & Education* 23, no. 1 (1996): 38–45.

Murphy, Alicia. *A Philosophy of the Use of Technology in Ministry*. Canadian Centre for Worship Studies. CCWS, 2003.

Nelson, David P. "Voicing God's Praise: The Use of Music in Worship." In *Authentic Worship: Hearing Scripture's Voice, Applying Its Truths*, edited by Herbert Bateman, 145–70. Grand Rapids: Kregel Publications, 2002.

Nelson, Tom. "Gifted for Work." In *Work Matters: Connecting Sunday Worship to Monday Work*, 143–62. Wheaton: Crossway, 2011.

Neufeld, Thomas R. Yoder. *Ephesians*. Believers Church Bible Commentary. Scottsdale: Herald Press, 2001.

Nicolosi, Barbara. "The Artist: What Exactly Is an Artist, and How Do We Shepherd Them?" In *For the Beauty of the Church*, edited by W. David O. Taylor, 103–21. Grand Rapids: Baker Books, 2010.

Noth, Personennamen. "Oholiab." In *Encyclopedia Judaica*, edited by F. Skolnik and M. Berenbaum, 393–94. Farmington Hills: Keter Publishing House, Ltd., 2007.

Nunes, Joseph C. and Andrea Ordanini. "I Like the Way It Sounds: The Influence on a Pop Song's Place in the Charts." *Musicae Scientiae* 18, no. 4 (2014): 392–409.

O'Brien, Peter Thomas. "Colossians, Letter to the." In *Dictionary of Paul and His Letters*, edited by Gerald F. Hawthorne, Ralph P. Martin, and Daniel G. Reid, 147–52. Downers Grove: InterVarsity Press, 1993.

-----. "Colossians." In *New Bible Commentary*, edited by D. A. Carson, R. T. France, J. A. Motyer, and G. J. Wenham, 1274. Downers Grove: Inter-Varsity Press, 1994.

-----. *Letter to the Hebrews*. Grand Rapids: W. B. Eerdmans, 2010.

O'Connor, Michael. "The Singing of Jesus." In *Resonant Witness: Conversations between Music and Theology*, edited by Jeremy Begbie and Steven R. Guthrie, 434–53. Grand Rapids: Wm B. Eerdmans Publishing, 2011.

Olford, Stephen F. and David L. Olford. "The Preacher and Music." *Anointed Expository Preaching*. Nashville: B&H Publishers, 1998.

Oliveros, Pauline. "My 'American Music': Soundscape, Politics, Technology, Community."

American Music 25, no. 4 (2007): 389–404.

Osborne, Grant R. *The Hermeneutical Spiral: A Comprehensive Introduction to Biblical Interpretation.* Downers Grove: InterVarsity Press, 2006.

-----. *Colossians & Philemon: Verse by Verse.* Bellingham: Lexham Press, 2016.

Otto, Christ John. *Bezalel: Redeeming a Renegade Creation.* Boston: Belonging House Creative, 2015, Kindle.

Pao, David W. *Colossians and Philemon.* Zondervan Exegetical Commentary on the New Testament. Grand Rapids: Zondervan, 2012.

Partridge, Michael. "Performing Faiths – Patterns, Pluralities and Problems in the Lives of Religious Traditions." In *Faithful Performances: The Enactment of Christian Identity in Theology and the Arts,* edited by Trevor Hart and Steven R. Guthrie, 75–89. Burlington: Ashgate Publishing Group, 2007

Pearson, Calvin. "'Lifting Holy Hands': Nuance, Nuisance, or Error? A Biblical Theology of the Practice of Lifting Hands in Worship." *The Artistic Theologian* 2 (2013): 26–36.

Peneycad, John. *Worship and the Screen.* Canadian Centre for Worship Studies. CCWS, 2006.

Peterman, Gerald. "Colossians." In *The Moody Bible Commentary,* edited by Michael Rydelnik and Michael G. Vanlaningham, 1867–75. Chicago: Moody Publishers, 2014.

-----. "Ephesians." In *The Moody Bible Commentary,* edited by Michael Rydelnik and Michael G. Vanlaningham, 1845–56. Chicago: Moody Publishers, 2014.

Peterson, Eugene. "The Pastor: How Artists Shape Pastoral Identity." In *For the Beauty of the Church,* edited by W. David O. Taylor, 83–101. Grand Rapids: Baker Books, 2010.

Porter, W. J. "Music." In *Dictionary of New Testament Background: A Compendium of Contemporary Biblical Scholarship,* edited by S. E. Porter and C. A. Evans, 711–19. Downers Grove: InterVarsity Press, 2000.

Quash, B. "Real Enactment: The Role of Drama in the Theology of Hans Urs Balthasar." In *Faithful Performances: The Enactment of Christian Identity in Theology and the Arts,* edited by Trevor Hart and Steven R. Guthrie, 13–32. Burlington: Ashgate Publishing Group, 2007.

Reasoner, Mark. "Purity and Impurity." In *Dictionary of Paul and His Letters,* edited by Gerald F. Hawthorne, Ralph P. Martin, and Daniel G. Reid, 775–76. Downers Grove: InterVarsity Press, 1993.

Redman, Robb. "Worship Wars or Worship Awakening?" *Liturgy* 19, no. 4 (2004): 39–44.

Reumann, John H. "Psalm 22 at the Cross: Lament and Thanksgiving for Jesus Christ." *Interpretation: A Journal of Bible and Theology* 28, no. 1 (1974): 39–58.

Reymond, Bernard. "Music and Practical Theology." *International Journal of Practical Theology* 5, no. 1 (2001): 82–93.

Rienstra, Ron. "Audio Technology in Worship: Keeping the Central Things Central." *Cross Accent* 21, no. 3 (2013): 26–30.

Risbridger, John. *The Message of Worship.* Downers Grove: InterVarsity Press, 2015.

Roberts, Mikie Anthony. "Hymnody and Identity: Congregational Singing as a Construct of Christian Community Identity." PhD dissertation, University of Birmingham, 2014.

Rothkoff, Aaron. "Tabernacle." In *Encyclopedia Judaica,* edited by Fred Skolnik and Michael Berenbaum, 418–24 Farmington Hills: Keter Publishing House, Ltd., 2007.

Rowan, Randy H. "Contemporary Worship as a Tool for Deepening Baby Boomers' Spiritual Well-Being." DMin dissertation, Asbury Theological Seminary, 2000.

Rumpf, Oscar J. "An Audio-Visual." *Journal of Bible and Religion* 31, no. 4 (1963): 329–31.

Ruth, Lester. "Don't Lose the Trinity! A Plea to Songwriters." The Robert E. Webber Institute for Worship Studies. 2006.

Michael Rydelnik and James Spencer "Isaiah." In *The Moody Bible Commentary,* edited by Michael

Rydelnik and Michael G. Vanlaningham, 1005–102. Chicago: Moody Publishers, 2014.

Ryden, Ernest Edwin. *The Story of Our Hymns*. Rock Island: Augustana Book Concern, 1931.

Ryken, Leland. *The Liberated Imagination: Thinking Christianly about the Arts*. Wheaton: H. Shaw Publishers, 1989.

Ryken, Philip Graham. *Exodus: Saved for God's Glory*. Wheaton: Crossway, 2005.

-----. *Art for God's Sake*. Phillipsburg: P&R Publishing, 2006, Kindle.

Sarna, Nahum M. *The JPS Torah Commentary: Exodus*. Philadelphia: Jewish Publication Society, 1991.

-----. *Exploring Exodus: The Origins of Biblical Israel*. New York: Schocken Books, 1996.

Sasser, Samuel Lee. "Worship: An Encounter with the Living God." Master's thesis, Fuller Theological Seminary, 1996.

Sauer, Ronald. "Hebrews" In *The Moody Bible Commentary*, edited by Michael Rydelnik and Michael G. Vanlaningham, 1921–46. Chicago: Moody Publishers, 2014.

Schaeffer, Francis A. "Art in the Bible." *The Complete Works of Francis A. Schaeffer: Vol. 2*, 375–92. Westchester: Crossway Books, 1982.

Schmit, Clayton J. "Technology and Art in Worship and Preaching." *Living Pulpit* 12, no. 3 (2003) 40–41.

Schultze, Quentin J. "Questions About Worship and Technology: A Starting Point for Discussion." *Reformed Worship* 65 (2002): 44.

-----. *High-Tech Worship?: Using Presentational Technologies Wisely*. Grand Rapids: Baker Books, 2004.

Schultze, Quentin J., D. J. Chuang, and Robb Redman. "Worship, Technology, and the Church: A Discussion with Quentin Schultze and DJ Chuang." *Cultural Encounters* 1 (2012): 93–104.

Schwarzer, Mitchell. "The Architecture of Talmud." *Journal of the Society of Architectural Historians* 60, no. 4 (2001): 474–87.

Sherry, Patrick. *Spirit and Beauty: An Introduction to Theological Aesthetics*. New York: Clarendon Press, 1992.

-----. "The Arts of Redemption." In *Faithful Performances: The Enactment of Christian Identity in Theology and the Arts*, edited by Trevor Hart and Steven R. Guthrie, 189–98. Burlington: Ashgate Publishing Group, 2007.

-----. "Beauty and Divinity." In *The Oxford Handbook of Religion and Arts*, edited by Frank Burch Brown, 44–56. New York: Oxford University Press, 2014.

Sigler, R. Matthew. "Not Your Mother's Contemporary Worship: Exploring CCLI's 'Top 25' Lists for Changes in Evangelical Contemporary Worship." *Worship* 87, no. 5 (2013): 445–63.

Slemming, Charles W. *Made According to Pattern: A Study of the Tabernacle in the Wilderness*. Fort Washington: Christian Literature Crusade, 1993.

Smith, Chuck. *The History of Calvary Chapel*. Costa Mesa: Calvary Chapel, 1981.

Smith, Gary V. *Isaiah 1–39*. NAC. Nashville: B&H Publishing, 2007.

Smith, J. A. "The Ancient Synagogue, the Early Church and Singing." *Music & Letters* 65, no. 1 (1984): 1–16.

-----. "First-Century Christian Singing and Its Relationship to Contemporary Jewish Religious Song." *Music & Letters* 75, no. 1 (1994): 1–15.

Smith, James E. *The Pentateuch*. Old Testament Survey Series. Joplin: College Press Pub. Co., 1992.

Smith, Jeremy. *Rebuilding: A Nineteen Week Devotional for those Serving in Church Technology*. USA: ChurchMag Press, 2016.

Smuts, Aaron. "The Power to Make Others Worship." *Religious Studies* 48, no. 2 (2012): 221–37.

Soanes, Catherine and Angus Stevenson. *Concise Oxford English Dictionary*. Oxford: Oxford University Press, 2004.

Sommer, Benjamin. "Conflicting Constructions of Divine Presence in the Priestly Tabernacle." *Biblical Interpretation* 9, no. 1 (2001): 41–63.

Spence-Jones, H. D. M. *The Pulpit Commentary: Colossians*. New York: Funk & Wagnalls Company, 1909.

-----. *The Pulpit Commentary: Exodus (Vol 2.)*. New York: Funk & Wagnalls Company, 1909.

Spinks, Bryan D. "Entertaining Worship or Worship as Entertainment? Megachurch Seeker Services and Multi-Sensory Worship" In *The Worship Mall: Contemporary Responses to Contemporary Culture*, 63–90. London: SPCK Publishing, 2010.

Sproul, R. C. *How Then Shall We Worship?: Biblical Principles to Guide Us Today*. Colorado Springs: David C. Cook, 2013.

Spurgeon, Charles H. "Psalm 22." In *The Treasury of David*. Romans45.org, 1869.

-----. *Lectures to My Students: A Selection from Addresses Delivered to the Students of the Pastors' College, Metropolitan Tabernacle*. London: Passmore and Alabaster, 1875.

Stamoolis, James. "Scripture and Hermeneutics: Reflections over 30 Years." *Evangelical Review of Theology* 28, no. 4 (2004).

Standhartinger, Angela. "Colossians and the Pauline School." *New Testament Studies* 50, no. 4 (2004), 572–93.

Stearns, Michelle. "Participation." In *It Was Good: Making Music to the Glory of God*, edited by Ned Bustard, 3427–710. Baltimore: Square Halo Books, 2013, Kindle.

Stein, Israel C. "Sacred Space and Holy Time." *Jewish Bible Quarterly* 34, no. 4 (2006): 244–46.

Stevens, R. Paul. "Spirit Work – Bezalel." In *Work Matters: Lessons from Scripture*, 40–48. Grand Rapids: Wm. B. Eerdmans, 2012.

Stiekes, Gregory J. "Liturgy in the Pastoral Epistles." *The Artistic Theologian* 2 (2013): 37–50.

Still, Todd D. "Christos as Pistos: The Faith(fulness) of Jesus in the Epistle to the Hebrews." In *Cloud of Witnesses: The Theology of Hebrews in Its Ancient Contexts*, edited by Richard Bauckham, Daniel Driver, Trevor Hart, and Nathan MacDonald, 40–50. New York: T&T Clark, 2008.

Stocker, David. "Hallelujah! Prayer-and-praise Worship and Formal Worship: Some Personal Reflections about Essence, Intent, and Blended Worship." *The Choral Journal 49*, no. 2 (2008): 67–71.

Stolzfus, Philip E. *Theology as Performance: Music, Aesthetics, and God in Modern Theology*. New York: T&T Clark, 2006.

Stowe, David. *No Sympathy for the Devil: Christian Pop Music and the Transformation of American Evangelicalism*. Chapel Hill: University of North Carolina Press, 2011.

Strawbridge, Greg. "Instruments." In *It Was Good: Making Music to the Glory of God*, edited by Ned Bustard, 5225–593. Baltimore: Square Halo Books, 2013.

Strickland, Diane. "Trinity Matters: The Trinity Needs to Be Named Regularly in Our Worship." *Reformed Worship* 51, no. 3 (1999): 9.

Stringer, Martin. *A Sociological History of Christian Worship*. New York: Cambridge University Press, 2005.

Strong, James. *The Tabernacle of Israel in the Desert*. Providence: Harris, Jones & Co., 1888.

-----. "The Tabernacle." *The Biblical World* 1, no. 4 (1893): 270–77.

Stuart, Douglas K. *Exodus*. NAC. Nashville: B&H Publishers, 2006.

Talbert, Charles H. *Ephesians and Colossians*. Grand Rapids: Baker Academic, 2007.

Taylor, Richard. *How to Read a Church: A Guide to Symbols and Images in Churches and Cathedrals.* Mahawh: HiddenSpring, 2005.

Taylor, W. David O. "Discipling the Eyes: The Visual Arts Can Play a Powerful Role in Worship—If We Look Closely Enough." *Christianity Today* 56, no. 4 (2012): 40–43.

Thiessen, Gesa Elisbeth. "Artistic Imagination and Religious Faith." In *The Oxford Handbook of Religion and the Arts,* edited by Frank Burch Brown, 77–90. New York: Oxford University Press, 2014.

Thiselton, Anthony. "Wisdom in the Jewish and Christian Scriptures: Wisdom in the New Testament." *Theology* 114, no. 4 (2011): 260–68.

Thomas, Robert L. *New American Standard Hebrew-Aramaic and Greek Dictionaries: Updated Edition.* Anaheim: Foundation Publications, Inc., 1998.

Timm, Eric Samuel. *Static Jedi: The Art of Hearing God Through the Noise.* Lake Mary: Charisma House, 2013.

Toledo, David M. "Why Worship Leaders Should Study Theology." *The Artistic Theologian* 2 (2013): 17–25.

Torrance, James B. "The Place of Jesus Christ in Worship." In *Theological Foundations for Ministry,* edited by Ray S. Anderson. Grand Rapids: Wm. B. Eerdmans (1979): 348–69.

-----. *Worship, Community & the Triune God of Grace.* Downers Grove: InterVarsity Press, 1996.

Tozer, A. W. and James L. Snyder. *Tozer on Worship and Entertainment.* Camp Hill: Wingspread, 1997.

Turner, Max. "Ephesians." In *New Bible Commentary,* edited by D. A. Carson, R. T. France, J. A. Motyer, and G. J. Wenham, 1242. Downers Grove: Inter-Varsity Press, 1994.

-----. "Spiritual Gifts and Spiritual Formation in 1 Corinthians and Ephesians." *Journal of Pentecostal Theology* 22 (2013): 187–205.

Utley, Robert James. *Paul Bound, the Gospel Unbound: Letters from Prison.* Marshall: Bible Lessons International, 1997.

-----. *The Superiority of the New Covenant: Hebrews.* Marshall: Bible Lessons International, 1999.

Uyen, Paul Cau. "Faith, Imagination, and Discipleship in Hebrews: The Role of Imagination in Faith for Discipleship." Master's thesis, Wheaton College, 2012.

VanAntwerp, Jeremy G. "The Person and Work of the Holy Spirit in Engineering." 2008 Christian Engineering Educators Conference. Beaver Falls: Geneva College, 2008.

Van Beek, Joanne. *Worship as Response to a Revelation of the Christ-Event Story.* Canadian Centre for Worship Studies. CCWS, 2004.

Van Dam, C. "Golden Calf." In *Dictionary of the Old Testament: Pentateuch,* edited by T. Desmond Alexander and David W. Baker, 368–72. Downers Grove: InterVarsity Press, 2003.

-----. "Priestly Clothing." In *Dictionary of the Old Testament: Pentateuch,* edited by T. Desmond Alexander and David W. Baker, 643–46. Downers Grove: InterVarsity Press, 2003.

Van Duzer, Jeff, Randal S. Franz, Gary L. Karns, Kenman L. Wong, and Denise Daniels. "It's Not Your Business: A Christian Reflection on Stewardship and Business." *Journal of Management, Spirituality & Religion* 3, no. 4 (2006): 348–74.

Van Voolen, Edward. "Judaism—Visual Art and Architecture." In *The Oxford Handbook of Religion and Art,* edited by Frank Burch Brown, 270–78. New York: Oxford University Press, 2014.

Vanhoozer, Kevin J. "Praising in Song: Beauty and the Arts." In *The Blackwell Companion to Christian Ethics,* edited by Stanley Hauerwas and Samuel Wells, 110–22. Malden: Blackwell Publishing Ltd, 2004.

-----. *The Drama of Doctrine: A Canonical-Linguistic Approach to Christian Theology.* Louisville: Westminster John Knox Press, 2005.

-----. "Forming the Performers: How Christians Can Use Canon Sense to Bring Us to Our (Theodramatic) Senses." *Edification: The Transdisciplinary Journal of Christian Psychology* 4, no. 1 (2010): 5–16.

-----. *Faith Speaking Understanding: Performing the Drama of Doctrine.* Louisville: Westminster John Knox Press, 2014.

Veith, Gene Edward. *The Gift of Art: The Place of the Arts in Scripture.* Downers Grove: Inter-Varsity Press, 1983.

-----. *State of the Arts: From Bezalel to Mapplethorpe.* Wheaton: Crossway, 1991.

-----. "Stealing Beauty." *World Magazine*, world.wng.org, March 20, 2004.

-----. "Vocation: The Theology of the Christian Life." *Journal of Markets & Morality* 14, no. 1 (2011): 119–31.

Viladesau, Richard. *Theology and the Arts: Encountering God through Music, Art, and Rhetoric.* New York: Paulist Press, 2000.

-----. "Aesthetics and Religion." In *The Oxford Handbook of Religion and the Arts*, edited by Frank Burch Brown, 25–43. New York: Oxford University Press, 2014.

Viljoen, Francois P. "Song and Music in the Pauline Epistles: Paul's Utilization of Jewish, Roman, and Greek Musical Traditions to Encourage the Early Christian Communities to Praise God." In *Die Skriflig* 35, no. 3 (2001): 423–42.

Volf, Miroslav. "Human Work, Divine Spirit, and New Creation: Toward a Pneumatological Understanding of Work." *The Journal of the Society for Pentecostal Studies* (Fall 1987): 173–93.

-----. *Work in the Spirit: Toward a Theology of Work.* Eugene: Wipf and Stock Publishers, 2001.

-----. "Reflections on a Christian Way of Being-in-the-World." In *Worship: Adoration and Action*, edited by D. A. Carson. Eugene: Wipf and Stock Publishers, 2002.

Wallace, Nathaniel. "Cultural Process in the Iliad 18:478–608, 19:373–80 ('Shield of Achilles') and Exodus 25:1–40:38 ('Ark of the Covenant')." *College Literature* 35, no. 4 (2008): 55–74.

Walvoord, John F. "Incarnation of the Son of God." *Bibliotheca Sacra* 105, no. 417 (1948): 145–53.

Warren, Rick. *The Purpose Driven Church: Every Church Is Big in God's Eyes.* Grand Rapids: Zondervan, 1995.

Watson, Francis. "Theology and Music." *Scottish Journal of Theology* 51 (1998): 435–63.

Weber, Tanya. "Art Education Needs No Justification!" *Journal of Education & Christian Belief* 8, no. 2 (2004): 113–28.

Wechsler, Michael L. "Psalm 22: A Prophetic Perspective on the Crucifixion of the Messiah" In *The Moody Bible Commentary*, edited by Michael Rydelnik and Michael G. Vanlaningham, 778–80. Chicago: Moody Publishers, 2014.

Wellhausen, Julius. *Prolegomena to the History of Israel*, translated by M. A. Black and B. D. Menzies. Oxford: Project Gutenberg Literary Archive Foundation, 1885, Kindle.

Werner, Eric. "If I Speak in the Tongues of Men...: St. Paul's Attitude toward Music." *Journal of the American Musicological Society* 13, no. 1 (1960): 18–23.

Westcott, Brooke Foss. *The Epistle to the Hebrews: The Greek Text with Notes and Essays.* Classic Commentaries on the Greek New Testament. London: Macmillan, 1903.

Westermeyer, Paul. "Christianity and Music." In *The Oxford Handbook of Religion and the Arts*, edited by Frank Burch Brown, 286–93. New York: Oxford University Press, 2014.

White, Kevin. "Drop the Mic." *First Things* 328 (2012): 19–21.

Whitley, Charles Francis. "The Language and Exegesis of Isaiah 8:16–23." *Zeitschrift Für Die Alttestamentliche Wissenschaft* 90, no. 1 (1978): 28–43.

Williams, Kesha Morant and Omotayo O. Banjo. "From Where We Stand: Exploring Christian Listeners' Social Location and Christian Music Listening." *Journal of Media and Religion*

12, no. 4 (2013): 196–216.

Willimon, William H. *The Bible: A Sustaining Presence in Worship.* Valley Forge: Judson Press, 1981.

Willis, Timothy M. "May You Be Like Bezalel!" *Restoration Quarterly* 57, no. 2 (2015): 109–14.

Wilson, Len and Jason Moore. *The Wired Church 2.0.* Nashville: Abingdon Press, 2008.

Wilson, Walter T. "The 'Practical' Achievement of Colossians: A Theological Assessment." *Horizons in Biblical Theology* 20, no. 1 (1998): 49–74.

Witherington, Ben. "Exposition and Exhortation, Round Two: Lower than the Angels, Greater than Moses, Gone through the Heavens, Entering His Rest." In *Letters and Homilies for Jewish Christians: A Socio-Rhetorical Commentary on Hebrews, James and Jude*, 139–61. Downers Grove: IVP Academic, 2007.

Witvliet, John D. "The Opening of Worship: Trinity." In *A More Profound Alleluia: Theology and Worship in Harmony*, edited by Leanne Van Dyk, 1–30. Grand Rapids: W.B. Eerdmans, 2005.

-----. "The Trinitarian DNA of Christian Worship: Perennial Themes in Recent Theological Literature." *Colloquium Journal* 2 (2005): 1–22.

-----. "The Joy of Christ-centered, Trinitarian Worship." *Worship Leader Magazine*, worshipleader.com, 2011.

-----. "Sustaining Pastoral Excellence: Pastoral Excellence and Christian Worship." *Faith and Leadership*. faithandleadership.com, 2015.

Wolff, Richard F. "A Phenomenological Study of In-Church and Televised Worship." *Journal for the Scientific Study of Religion* 38, no. 2 (1999): 219–35.

Wright, Christopher. *Knowing the Holy Spirit Through the Old Testament.* Downers Grove: IVP Academic, 2006.

Wright, Jay. *Three Essentials of Biblical Worship.* Canadian Centre for Worship Studies. CCWS, 2006.

Wright, N. T. *Colossians & Philemon: 8 Studies for Individuals and Groups.* Downers Grove: InterVarsity Press, 2009.

-----. *Ephesians: 11 Studies for Individuals and Groups.* Downers Grove: InterVarsity Press, 2009.

-----. *Hebrews for Everyone.* Louisville: Westminster John Knox Press, 2004.

-----. *Paul for Everyone: The Prison Letters.* Louisville: Westminster John Knox Press, 2002.

Zschech, Darlene. *Extravagant Worship.* Bloomington: Bethany House Publishers, 2002.

Zschomler, Gregory E. *Lights, Camera, Worship!: Redefining Media and Technical Arts for Today's Worship.* Vancouver: Eyrie Press, 2014.

ABOUT THE AUTHOR

Josiah Way (PhD, University of Birmingham UK) is the Director of Multimedia Services at California Baptist University in Riverside, CA, and serves as a regional Technical Director at Saddleback Church the Aliso Viejo campus. With over 25 years' experience in pro-AV, working in live production, TV, film, studio recording, sports, theatre, and church tech, he holds a PhD in Modern Theology from the University of Birmingham UK, Master's in Applied Biblical Studies from Moody Theological Seminary, Bachelor's in Philosophy from the University of Southern California, and AVIXA CTS Certification. Joe serves on multiple boards in both the higher education and pro-AV industries, speaks and consults regularly worldwide, hosts the *Higher Ed AV* podcast, and is a contributing writer for various trade publications including *Church Production Magazine*. He and his wife, Amy, reside in Lake Forest, CA, and have three children between them.

Joe is available for speaking, course instruction, and consulting. Contact at: joe@josiahway.com or www.josiahway.com.

More information on this book and supplemental materials are available at: www.producingworship.com.

48189818R00124

Made in the USA
Middletown, DE
12 June 2019